Solving the Evolutionary Puzzle of Human Cooperation

Scientific Studies of Religion: Inquiry and Explanation

Series editors: Luther H. Martin, Donald Wiebe, William W. McCorkle Jr., D. Jason Slone and Radek Kundt

Scientific Studies of Religion: Inquiry and Explanation publishes cutting-edge research in the new and growing field of scientific studies in religion. Its aim is to publish empirical, experimental, historical and ethnographic research on religious thought, behaviour and institutional structures. The series works with a broad notion of scientific that includes innovative work on understanding religion(s), both past and present. With an emphasis on the cognitive science of religion, the series includes complementary approaches to the study of religion, such as psychology and computer modelling of religious data. Titles seek to provide explanatory accounts for the religious behaviours under review, both past and present.

The Attraction of Religion, edited by D. Jason Slone and James A. Van Slyke
The Cognitive Science of Religion, Edited by D. Jason Slone and William W. McCorkle Jr.
Contemporary Evolutionary Theories of Culture and the Study of Religion, Radek Kundt
Death Anxiety and Religious Belief, Jonathan Jong and Jamin Halberstadt
Language, Cognition, and Biblical Exegesis, edited by Ronit Nikolsky, Istvan Czachesz, Frederick S. Tappenden and Tamas Biro
The Mind of Mithraists, Luther H. Martin
New Patterns for Comparative Religion, William E. Paden
Philosophical Foundations of the Cognitive Science of Religion, Robert N. McCauley with E. Thomas Lawson
Religion Explained?, edited by Luther H. Martin and Donald Wiebe
Religion in Science Fiction, Steven Hrotic
Religious Evolution and the Axial Age, Stephen K. Sanderson
The Roman Mithras Cult, Olympia Panagiotidou with Roger Beck

Solving the Evolutionary Puzzle of Human Cooperation

Glenn Barenthin

BLOOMSBURY ACADEMIC
LONDON • NEW YORK • OXFORD • NEW DELHI • SYDNEY

Bloomsbury Publishing Plc
50 Bedford Square, London, WC1B 3DP, UK
1385 Broadway, New York, NY 10018, USA

BLOOMSBURY, BLOOMSBURY ACADEMIC and the Diana logo are trademarks of
Bloomsbury Publishing Plc

First published in Great Britain 2020

Copyright © Glenn Barenthin, 2020

Glenn Barenthin has asserted his right under the Copyright, Designs and
Patents Act, 1988, to be identified as Author of this work.

Cover image © Shutterstock

All rights reserved. No part of this publication may be reproduced or transmitted
in any form or by any means, electronic or mechanical, including photocopying,
recording, or any information storage or retrieval system, without prior
permission in writing from the publishers.

Bloomsbury Publishing Plc does not have any control over, or responsibility for,
any third-party websites referred to or in this book. All internet addresses given in this
book were correct at the time of going to press. The author and publisher regret any
inconvenience caused if addresses have changed or sites have ceased to exist, but
can accept no responsibility for any such changes.

A catalogue record for this book is available from the British Library.

A catalog record for this book is available from the Library of Congress.

ISBN:	HB:	978-1-3501-0675-8
	ePDF:	978-1-3501-0676-5
	eBook:	978-1-3501-0677-2

Series: Scientific Studies of Religion: Inquiry and Explanation

Typeset by Integra Software Services Pvt. Ltd.
Printed and bound in Great Britain

To find out more about our authors and books visit www.bloomsbury.com
and sign up for our newsletters.

For Lisa, Rachel, Elliott and Curtis

Contents

List of figures	x
Foreword	xi
Acknowledgements	xiii
List of abbreviations	xiv
Introduction	1
1 Minds, gods and group selection theory	17
2 A modest proposal	57
3 Family matters	93
4 From 'thin' to 'thick cooperation'	131
5 How does that make sense?	169
6 The road to 'Denmark'	199
References	209
Index	227

Figures

1 Wason Four Card Selection Task 23
2 Memory Test 64

Foreword

The cognitive science of religion (CSR), as I have argued elsewhere, is a progressive rather than a degenerating research programme that not only has developed a standard model but also is inferentially rich and capable, therefore, of developing in new directions. Glenn Barenthin in this excellent book explores a new road in the wide terrain of the CSR. He leads the way towards a deeper understanding of human cooperation, a subject of compelling interest in the CSR, and by doing so clarifies issues about the role of the gods in human thought and behaviour.

One of the many problems that cognitive scientists of religion are interested in solving is accounting for cooperative behaviour given the 'selfish' way in which evolution proceeds. Why would people cooperate at the expense of their genetic fitness? Kin selection and direct and indirect reciprocity have been proposed as solutions. These solutions work in the kinds of small groups typical of the early stages of human evolution but seem more problematic in explaining human behaviour when larger social systems are the name of the game. Of special importance is the role that free-riders play when cooperation does occur. Something more than either depending on the possibility of reciprocation or ensuring a future for kin seems to be required to keep free-riders in check. An intriguing proposal has emerged: moralizing gods, also referred to as Big Gods. Big Gods demand respect, induce fear and precipitate anxiety, or so it seems. Surely believing in such powerful beings, capable of eternal punishment, can function to keep free-riders in line!

Well, I won't spoil your fun by giving you Barenthin's answer, but read this book and you will find out his solution and a great deal more besides, for example about the role that institutions play in social life.

In the initial stages of the development of the CSR, the focus was, by necessity, on the cognitive mechanisms that gave rise to religious notions and the cognitive processes at work. Attention was paid to agency detection, memory dynamics, action representation, theory of mind and so on. But at some stage of conceptual development in the research programme, it became obvious that the existence of these mechanisms also required an explanation, and here evolutionary psychology provided the scientific wherewithal that cognitive scientists interested in religion could examine and employ. Cognitive scientists began to explore the role that natural selection played in designing these cognitive capacities. Barenthin has taken a close look at this very important development. He has examined the theoretical proposals, evaluated the empirical and experimental work and arrived at some significant insights. As a result, his book represents an important contribution to our understanding of the roots of human social behaviour. You certainly will discover why it is interesting to know whether moral systems require the notion of gods in order to be successful, and, frankly, you will know more about evolution by reading this book and taking its arguments into consideration. Enjoy a great read.

E Thomas Lawson
Queen's University Belfast

Acknowledgements

I wish to express my sincere gratitude to the editors of this series, Luther Martin and Don Wiebe, for their tremendous support during the production of this book. I have known Don for a number of years, and I am profoundly grateful for his guidance, encouragement and generous support throughout my academic life at Trinity College. His selfless devotion to not only myself but also all students is an exceptional characteristic and a great example for university administrators to emulate. Don, more than anyone, has taught me how to read carefully and think critically.

I would like to thank David Neelands for his enthusiastic optimism and his willingness to step in at a moment's notice. I have learned a great deal from Charles Helwig, Marsha Hewitt, Ivan Khan and Morris Moscovitch. The conversations we shared were invaluable and extremely helpful in my research. During my studies, I have had the pleasure of conversations with many who, in their own way, have left a positive mark on my thinking.

I would also like to thank Tom Lawson for his kind words and his overwhelming support. It was not until I read Tom's book, *Rethinking Religion: Connecting Cognition and Culture*, that I started to think seriously about the human mind and religion. Recently, I had the opportunity to meet Tom, and from that brief encounter, I can understand why Robert McCauley describes Tom as a 'wondrous human being'.

There are many colleagues, friends and students who have inspired and challenged my thoughts on an array of topics. I will single out John Irwin, an exemplar of cooperation, for his thought-provoking conversations to and from GH.

A special thank you to Helen, my mother-in-law, for her unconditional support. A very special thank you to my daughter Rachel and sons Elliott and Curtis for their tremendous support and for our dinner conversations that helped clear up some of my murky thinking. Above all, I want to thank my wife Lisa, who is my greatest fan, editor, critic and best friend. I am indebted for all her support, not only for this book, but for everything. Without her selfless help, encouragement and love, I would not have been able to complete this book.

Abbreviations

ACS	Animal Communication Systems
ASC	Altered State of Consciousness
AVP	Arginine Vasopressin
BCE	Before the Common Era
BGP	Big God Proponents
CEDAW	Convention on the Elimination of All Forms of Discrimination Against Women
CRC	Convention on the Rights of the Child
CSR	Cognitive Science of Religion
DG	Dictator Game
DNA	Deoxyribonucleic Acid
ERP	Enhanced Religious Prime
HADD	Hypersensitive Agent Detection Device
IA	Inequity Aversion
LCA	Last Common Ancestor
NDE	Near-Death Experience
NPC	Neutral Prime Control
OXT	Oxytocin
PD	Prisoner's Dilemma
PGG	Public Good Games
RCT	Rational Choice Theory
REM	Rapid Eye Movement

SCCS	Standard Cross-Cultural Sample
SMH	Supernatural Monitoring Hypothesis
SPH	Supernatural Punishment Hypothesis
SRP	Standard Religious Prime
TASS	The Autonomous Set of Systems
ToM	Theory of Mind
UDHR	Universal Declaration of Human Rights

Introduction

In 1988, two friends, Jeremy Strohmeyer and David Cash Jr., were at a Nevada casino when Strohmeyer made eye contact with a seven-year-old girl. Strohmeyer took the girl into the male washroom, dragged her into a stall and sexually assaulted her. Cash was in the washroom as well and at one point went into the adjacent stall and hoisted himself up to witness the girl struggling and Strohmeyer muzzling her screams. Cash left without any attempt to intervene and met up with Strohmeyer a little later. Strohmeyer confessed to Cash that he had raped and brutally murdered the girl.

After Strohmeyer was arrested, the fact that Cash did nothing to help the young girl was made public and, understandably, this infuriated many people. To add insult to injury, Cash told the *LA Times* that he had done nothing wrong and indicated that he wasn't 'going to lose any sleep over somebody else's problems' (Booth 2001). He also confessed that he was more upset with losing his friend than having any concern for the brutal murder of the girl as he did not know the victim's family. Since there was no law at the time requiring witnesses to intervene in such cases, Cash was never charged with a criminal offence. Subsequent to the incident, legislators in the State of Nevada enacted the Sherrice law, making it mandatory for witnesses to report cases of sexual assault against children to the authorities. Fortunately, this law is only necessary for the most selfish individuals in society, as the vast majority of people would intervene if someone was in need of help.

Consider another incident that took place on 2 January 2007 at a subway station in Manhattan. Wesley Autrey, a fifty-year-old construction worker, was waiting for a train with his two young daughters Syshi and Shugui, when twenty-year-old Cameron Hollopeter had a seizure and collapsed to the ground. Autrey and two other women went to his aid. When Hollopeter got up, he stumbled and fell to the tracks, landing between two rails. When Mr. Autrey saw the train approaching, he passed his two daughters to one of the women and jumped down on to the tracks with the intention of pulling Hollopeter away. Realizing that he didn't have enough time to do what he had planned, Autrey laid on top of Hollopeter shielding him from the train above. The train came within inches of killing both men. If it was not for the bravery and quick action of Autrey, Hollopeter would certainly have been killed. When reporters asked him about the incident, Autrey humbly replied, 'I didn't feel like I did something spectacular, I just saw someone who needed help. I did what I felt was right' (Buckley 2007).

Naturally good or evil?

These two incidents offer extreme contrasts in the wide range of human behaviour. Of course, not everyone exhibits such courageous and selfless acts as Wesley Autrey, and fortunately society isn't plagued with too many individuals as pathetic as Cash. From these two conflicting examples, a question arises as to whether human beings are essentially good, in the sense that they care for others and will offer assistance, even at an expense to themselves, or are humans essentially selfish, like Cash, concerned only about their own well-being and not at all troubled by the problems of others?

Historically, philosophers have offered different accounts as to whether humans are naturally good or evil. Their understanding of human nature shaped their assessment on how to achieve greater cooperation and the necessary means to attain a less-violent world. Utopians, such as Jean-Jacques Rousseau, painted a simple, nonviolent paradise in the so-called state of nature, where humans were naturally peaceful with no desire to inflict harm on others (Rousseau 1762). According to Rousseau, natural human beings were solitary and rather than engage in fighting they would simply avoid confrontations. Rousseau voiced his displeasure with the ownership of property, as well as the problems associated with the corrupt Monarchy, and therefore thought human goodness was corrupted by civilization.

In contrast, Thomas Hobbes saw humans as basically nasty, and in order to have a healthy society, human selfish behaviour had to be suppressed. Like Rousseau, Hobbes imagined a state of nature but reasoned that each person had the right to preserve his own life. Since each man is out for himself, the state of nature is characterized as 'every man against every man' (Hobbes 1651). To escape from such an unstable state, human beings needed a strong central government or Leviathan in order to threaten and intimidate them into obeying the rules. With such a government, a state of barbarity would be avoided. For Hobbes, the road to Leviathan was either through violence, which he called 'commonwealth by acquisition', or a more peaceful route, coined 'commonwealth by institution', where men agreed to give up their power to a ruler or an assembly of men (Hobbes 1651).

In using the phrase 'state of nature', both Rousseau and Hobbes were addressing what they understood to be human nature. They were attempting to distinguish between what is human by nature and what is human by designed institutions. They took as their starting point a mythical time where individuals were isolated from each other, and they assumed that society or group living was not natural. Perhaps they saw the heuristic usefulness of fashioning their argument in such a manner; however, we now know that there was never a time when individual humans lived solitary lives. Humans do not come together as a result of a social contract; rather, living in a group is natural, and the different forms of cooperation are shaped and constrained by biology and culture. Nevertheless, having a better understanding of human nature is essential for comprehending the different forms of cooperation that have emerged in the evolution of *Homo sapiens*. The case to be made in this book is that human cooperation is not as puzzling as some researchers have claimed.

What is cooperation?

What does the term cooperation mean? Generally, the term refers to working together to achieve a goal or 'a form of working together in which one individual pays a cost (in terms of fitness, whether genetic or cultural) and another gains a benefit as a result' (Nowak and Coakley 2013, p. 4). These authors distinguish altruism from cooperation that simply happens by suggesting that altruism is motivated by specific goals or affective commitments. They go on to define altruism as 'a form of (costly) cooperation in which an individual is motivated by good will or love for another (or others)' (Nowak and Coakley 2013, p. 5). Clearly by this distinction, Wesley Autrey is altruistic.

By distinguishing cooperation and altruism in this manner, it may lead to a number of misconceptions. For instance, consider an individual who receives a phone call at 3:00 am from a friend who needs help because her car broke down. Is this type of helpful behaviour cooperation or altruism? Certainly this cooperative act is costly in the short term, but it may provide some long-term advantages in the future, so this kind act may be mutually beneficial rather than altruistic. Conceivably, there are a host of behaviours that can be described as altruistic but may involve a net direct fitness benefit for the actor. For this reason, I will include all altruistic acts as well as mutually beneficial behaviour in the definition of cooperation.

While cooperation comes in degrees, I argue that there are three distinct forms of cooperation: collaboration, 'thin cooperation' and 'thick cooperation'. Behaviour that is mutually collaborative involves two or more individuals pursuing a goal together that they would not be able to accomplish on their own. Collaboration is a type of cooperation that is found in the nonhuman world, and it can be achieved without the animal comprehending why they are behaving in such a fashion. This type of cooperation can be explained by biological evolution.

A second class of cooperation includes acts by an individual that benefit another, even though the individual is aware that it may incur costs. This includes such acts as helping and sharing with others. This 'thin cooperation' is distinguished from collaboration by the fact that it requires individuals to comprehend the reasons for their actions. Humans have the cognitive ability to evaluate the costs and benefits of their behaviour and act accordingly. This type of cooperation is only found in human societies, and it emerged from both biological evolution and cultural development. The term 'thin cooperation' does not necessitate that the particular society is small scale; rather, it pertains to the context of how cooperation is achieved. As the long argument in this book unfolds, I will show that human beings naturally engage in 'thin cooperation'.

I loosely adopt the distinction between 'thin' and 'thick cooperation' from anthropologist Clifford Geertz's conception of 'thin' and 'thick description', who in turn lifted the phrase from philosopher Gilbert Ryle (Geertz 1973). According to Geertz, what social anthropologists do is ethnography, and ethnography is, among other things, establishing rapport, interviewing informants and transcribing texts (Geertz 1973). However, these techniques and procedures are not what define the enterprise. For Geertz, what defines the undertaking is the type of intellectual effort that it is, and he suggests this enterprise is an elaborate venture in 'thick description'. For Geertz, 'thick description' is much more than simply data collection; rather, it describes behaviour in its *context*, as well as in such a way that it becomes meaningful to an outsider.

Similarly, 'thick cooperation' is more than simply describing a society as one where we find cooperation, but it also takes into account the context. To be classified as 'thick cooperation' as opposed to 'thin cooperation', attention must be focused on the individual's power of self-definition and consideration of the person's choice or freedom to act or not. 'Thick cooperation', at times, involves setting aside our natural predispositions to be with our own kind, as well as setting aside our individual and cultural biases and traditions. 'Thick cooperation' is an extension of 'thin cooperation' as it expands to include not only strangers, but also those who are commonly considered 'other' even within the in-group. 'Thick cooperation' requires the development of institutions that stress individual autonomy, and this type of cooperation can only be accomplished through a process of reasoning and argumentation and is therefore a cultural development.

One other word that may assist in distinguishing between 'thin' and 'thick cooperation' is coercion. As societies scaled up from our hunting and gathering forebears, the social order for most of recorded history was based on coercion by a leader and a coalition or network of elites. The leader and the elites cooperated with each other to ensure a system of production for their own interest, and they also cooperated with others who had the potential for challenging their privileges. Political scientists Douglas North, John Wallis and Barry Weingast call this model of social order the 'natural state'. The natural state can be described as 'thin cooperation', and it is the way that states have scaled up and down. In contrast, a social order that can be called 'thick cooperation' was not achieved until the seeds were planted for freedom, democracy and the recognition of individual autonomy.

The cooperation puzzle

Evolutionists since Darwin have questioned why humans engage in cooperative behaviour at a cost to their own genetic fitness (Axelrod 1984; Nowak 2011). Clearly, Wesley Autrey is a courageous, selfless human being, yet if human nature is 'red in tooth and claw', how did these prosocial tendencies evolve? Should we not all be more like David Cash? Historically, three evolutionary mechanisms have been suggested to account for the origin of human cooperation. William Hamilton advanced a genetic account of social behaviour with the idea of kin selection. This implies that people are inclined to assist others who are more genetically similar (Hamilton 1964). Second, Robert Trivers argued convincingly that genetically unrelated people can benefit from cooperation as long as they can count on receiving benefits from others in the future. This type of cooperation may initially be costly, but in the long run, the altruist will benefit through reciprocal altruism (Trivers 1971). For instance, you help a friend today and, inevitably, you will be rewarded for your benevolence in return. However, as Stuart West and colleagues point out, the term reciprocal altruism is confusing, and better alternatives include simply reciprocity or reciprocal cooperation (West et al. 2007b). For these authors, reciprocity is not altruistic because if the cooperative behaviour benefits both the actor and the recipient, then it is mutually beneficial even though it may appear to be altruistic. I will argue in Chapter 4 that while reciprocity played a large role in human cooperation, there is little evidence to suggest that it was a factor in nonhuman cooperation.

In addition to reciprocity, any act of kindness may become known to others, which would demonstrate that the person is helpful. This suggests that cooperation can be accounted for by cooperators advertising their good deeds in order to secure a favourable reputation that will encourage future help from others (Alexander 1987). Explaining cooperation for groups of friends or even small tribes can be accounted for by these three mechanisms – kin selection, direct reciprocity and indirect reciprocity. However, some scholars have suggested that these mechanisms leave unexplained the intensification of the free-rider problem, as groups scaled up from hunting and gathering tribes to large-scale societies (Wilson 2002; Norenzayan 2013). Accordingly, it has been suggested that there must be additional means to account for the cooperation puzzle (Nowak 2011). While I will argue that there is no need for such new mechanisms to account for the evolution of cooperation, I will briefly explain why researchers claim the need for more theoretical models.

Punishment

The free-rider problem rests on the assumption that people can get away with cheating. However, it would be reasonable to believe that people would evolve tendencies to identify cheaters and intervene when someone is not helping or contributing their fair share to the goals of the group. It was Robert Boyd and Peter Richerson who argued that as long as the cost of punishing was relatively low, if some members of the group punished free-riders, then this type of behaviour would emerge (Boyd and Richerson 1992). Punishment, they reasoned, was a powerful tool for group conformism. I will explore this idea in more detail in Chapters 2 and 3, but for now let's consider the problem of free-riders, especially in an expanding population. Since punishment is effective in small-scale society, as long as the cost is low, then what could be a cost-effective solution when the population increases? There is a group of researchers who suggest that it was the belief in a vigilant, moralizing God that was the crucial motivator for the prosocial behaviour necessary for large-scale cooperation.

The fear of God

The argument rests on the assumption that a fear of God or other supernatural agents discourages selfishness and moral transgressions. This implies a Hobbesian view of human nature and, according to this reasoning, without a stick to shape human behaviour, life would be nasty, brutish and short. These scholars maintain that there is clear evidence to suggest that a belief in punishing supernatural agents strongly influences cooperation, and such belief mobilized small human groups to expand into the civilizations we find today (Norenzayan 2013; Johnson 2016). One of the questions to be answered in this book is whether a belief in God was necessary for large-scale cooperation as these researchers argue? In order to answer this question, there will be a number of other questions that require attention first. For instance, is there adequate evidence to support the claim that a fear of god fosters moral or prosocial behaviour? Second, does it make sense historically to suggest that it was groups of god-fearing people who outcompeted groups of non-believing people, leading to the cooperation necessary for large-scale civilizations?

Psychologist Nicholas Humphrey has explored this relationship between a belief in God, or other supernatural forces, and an understanding of morality. Humphrey reasonably argues that even if people have not had a religious upbringing, they have all been exposed to the idea that supernatural agents watch over and care for them. He suggests that the deep connection between believing in supernatural agents and being a gracious and honest person has been so ingrained in society that if you are a non-believer, then you are considered generally wicked and not capable of moral virtue (Humphrey 1995). The polls taken across the United States of America seem to confirm Humphrey's observation.

According to a 2007 Gallup poll, most Americans in a presidential election campaign would readily vote for a Mormon, Jew or a homosexual over an atheist. In questions concerning the beliefs and behaviours of the American people, most ranked atheists below Muslims, homosexuals and immigrants in trustworthiness (Edgell 2006). When asked why they had such a negative view of atheists, the majority of people polled justified their scepticism by the assumed lack of morality of atheists. They believed that without a belief in God, one could not be moral. They went so far as to associate atheism with such things as drug abuse and prostitution, and they conceived anyone who did not believe in God as selfish and immoral (Edgell 2006). The information from these polls suggest that many people associate a belief in God with moral behaviour.

Factoid

Although it has been taken for granted that the idea of God promotes prosocial or moral behaviour, there has been very little evidence to support such a claim. American writer Norman Mailer coined the term 'factoid' to mean 'something that looks like a fact and could be a fact, but in fact is not a fact' (Pruden 2007). Richard Dawkins and Christopher Hitchens not only have argued that religion or a belief in God has very little to do with fact, but also contend that such a belief is harmful for society (Dawkins 2006; Hitchens 2007). In contrast, a group of researchers maintain that the idea of God has not only been good for humanity, but is also absolutely necessary, regardless of the truthfulness of the claims about God's existence. Some go so far as to suggest that the belief in God is what makes us human (Johnson 2016). These researchers insist that people are more likely to cooperate with strangers if they believe that they are being watched by morally, punishing gods or other supernatural agents. A leading advocate of this idea is Ara Norenzayan who, in *Big Gods: How Religion Transformed Cooperation and Conflict*, argued that, 'prosocial religions, with their Big Gods who watch, intervene and demand hard to fake loyalty displays, facilitated the rise of cooperation in large groups of anonymous strangers' (Norenzayan 2013). Another proponent of this hypothesis is Dominic Johnson, who invites us to believe that human behaviour is 'strongly influenced by the idea that God or a supernatural agent will reward or punish our actions' (Johnson 2016, p. 7). Johnson goes on to suggest that this was evolutionarily adaptive as 'the looming threat of supernatural punishment deterred selfish behaviour and increased cooperation and this was a good thing for individuals as well as society' (Johnson 2016, p. 8). Finally, he raises a concern that without a belief in supernatural punishers, we may be 'playing with fire' (Johnson 2016, p. 233).

While Norenzayan argues from a cultural group selection framework and Johnson argues that a belief in a moralizing, omniscient and omnipotent supernatural watcher was evolutionarily adaptive, both suggest that the idea of a mean God strongly reduces moral transgressions, causing greater successful group cohesion and thereby enabling societies to scale up. Without this type of 'belief-crutch', humans would resort to selfish, immoral behaviour (Kosslyn 2013). One major problem with these claims is the overly simplistic classification of distinct and discrete religions and the focus on the 'success' of religions with Big Gods. Historically, there has been significant division within all religions, questioning the cohesiveness of the group. For instance, most religions are divided by sects and subsects. Furthermore, the glorification of a threatening and moralizing God misses other compelling evidence and ideas that may influence the motivation of people. Partitioning people into religious groups and labelling some as prosocial, suggesting some are antisocial, is a poor way to understand human behaviour. These authors fail to take into account the internal diversities within these prosocial religions and the glaring fact that for most of history, half the population was never invited to cooperate.

Big God Proponents

Researchers, who I baptize the Big God Proponents (BGP), acknowledge that from the historical record one cannot unequivocally say that religious ideas are the fundamental ingredients necessary to behave prosocially (Norenzayan 2016). It could very well be that people who are prosocial are attracted to religious institutions, but the theologies of the different religions actually stifle cooperation by acting as boundary markers. Indeed, the prosocial religion hypothesis does not suggest that 'religion promotes indiscriminate sharing and caring, but rather the hypothesis that religion fosters social cohesion within religious groups' (McKay and Whitehouse 2016). Others have argued that while human prosocial behaviour is an evolved predisposition for small-scale group living, 'religions have from their social origins, been promoters of, perhaps the primary promoters of', what the authors refer to as 'assortative sociality' (Martin and Wiebe 2014, p. 3-4). By 'assortative sociality', these authors suggest that religious groups prioritize their differences from others. The bond that functions to unite the in-group also forms a barrier to anyone who is not considered a member of the group, thereby undermining the autonomy of the individual. Luther Martin and Don Wiebe point out that this may lead to 'ethnocentrism and Philopatry (reduced mobility outside one's natal group), intergroup vigilance and xenophobia' (Martin and Wiebe 2014, p. 4). Therefore, the authors advocating the fear of God hypothesis invariably ignore historical evidence and place a great deal of weight on psychological experiments in an attempt to support their claim that religious belief has a causal effect on prosocial tendencies (Johnson 2016; Norenzayan et al. 2016).

Solving the puzzle

At this point, I wish to make it clear that this book is not concerned with whether people who participate in religious activities are any kinder or more malicious than those who do not take part in such activities. Undoubtedly, the vast majority of people who attend churches, mosques, temples or synagogues are good people who would make kind neighbours, but one of the questions that needs to be answered is whether it is the idea of God that makes them good? In order to answer that, I will endeavour to determine whether people are the Hobbesian creatures depicted by the BGP to the extent that humans need to be threatened with death, or punishment after death, in order to cooperate with each other. For it also seems clear to me that the majority of people who do not partake in religious affairs are also kind and unselfish. So another question to be addressed in this book is, if prosocial tendencies do not come from a belief in a punishing God, then where do they come from? By proceeding in this manner, I suggest that it will provide us with the necessary clues for solving the cooperation puzzle. I will outline the format to be taken in this book presently.

Format

In Chapter 1, I assess the evidence offered by the Cognitive Science of Religion (CSR) researchers accounting for why people believe in gods. After all, if one of the questions to be answered in this book is whether the idea of God makes us nicer and more cooperative, then we should account for why people believe in gods in the first place. The evidence suggests that religion is an evolutionary by-product, as a belief in gods or supernatural agents emerged from features of human cognitive systems that evolved for comprehending the physical and social world (Lawson and McCauley 1990; Boyer 2001; Barrett 2004). Humans have a dual-processing cognitive system that includes brain mechanisms that are intuitive, meaning they are triggered automatically by both internal and external stimuli, and a second system that is characterized as slow, controlled by a central tive and is effortful (Stanovich 2004). According to CSR researchers, the in gods stems from our intuitive mind tools. How we come to think about the characteristics of the gods attractively fits with what people know about the nature of the world. Since humans are remarkably sensitive to signals of agency and intention, we are prone to believing in entities that may or may not exist. However, I will argue that the belief in gods is not simply the result of our intuitive mind tools, although that is the ultimate source. I also argue that once individual experiences are discussed, stories are created, and these stories are circulated and reinforced by the most influential people in the group.

While I find the evidence provided by the CSR researchers suggesting how humans came to believe in the gods persuasive, I question the evidence offered by the BGP in support of their claim that it was this belief in a moralizing, punishing God that provided the motivation for expansive prosocial behaviour. There are two different approaches to the BGP assertions. One suggests that belief in supernatural agents was a biological adaptation due to its beneficial effects on cooperation (Bering and Johnson 2005, 2007; Johnson 2016). A second group of BGP contends that although not a biological adaptation, belief in supernatural agents or gods is a functional adaptation at the level of the group (Wilson 2002, 2015; Norenzayan 2013). I will argue that the evidence supporting these arguments is weak, but in Chapter 1, I address the problems associated with cultural group selection in general and suggest that when it is extended to the argument that groups of people believing in gods outcompeted groups of non-believers, the argument does not survive a critical investigation.

Regardless of whether religion was a product of genetic or cultural evolution, or even a combination of both, the BGP claim that the enabling element that facilitated the rise of greater cooperation was a belief in punishing, moralizing supernatural agents. In order to support their thesis, the BGP reason that if a belief in supernatural agents causes prosocial tendencies, then they should be able to demonstrate experimentally that when individuals are induced with thoughts of God, they will demonstrate signs of increased prosocial behaviour. Much of the evidence that the BGP have to offer comes from a number of game theory experiments.

In Chapter 2, I will critically examine this evidence and assess whether there is any reason to believe that a person's belief in a punishing, moralizing God leads them to act prosocially. In many of the experiments, the BGP use the technique of priming. I argue that the priming literature in general is problematic. However, when it comes to priming people with religious words, the conclusions drawn from these experiments are even more open to doubt. One major problem is due to the fact that the religious words used are extremely ambiguous. For instance, the word 'spirit' is used in a number of experiments, and arguably 'spirit' could mean a number of different things to different people. Furthermore, I will show that there have been so many experiments using not just religious terms but all types of words that have led researchers to conclude that being primed with those words caused prosocial behaviour. These studies raise serious concerns as to how much weight should be placed on the interpretation from the priming studies. Finally, the BGP argument rests on the assumption that to dissuade people from moral transgressions, the harsher the penalty or threat of violence, the more effective it will be. I will provide evidence to question the accuracy of this claim.

While the BGP suggest that the idea of God was essential for binding groups together, I argue that humans naturally engage with others and are more likely to cooperate with kin, second friends and, with effort, strangers. In Chapter 3, I will show that the mammalian brain evolved to seek mates and to take care of others in a similar way that we take care of ourselves (Churchland 2013). I will argue that while all mammals have the preparedness for the helping behaviour for kin, it is humans alone who have the cognitive capacity to assess the costs and benefits of cooperative behaviour and therefore the capacity to modify their behaviour accordingly. This has given humans the capacity for more complex mechanisms to enforce cooperation such as laws, justice and trade that allowed for the rise of cooperative large-scale societies (Seabright 2004).

One of the main arguments in Chapter 3 is that for most of human history people lived in small kin groups, and this is where we must look to get a sense of our natural predispositions. By looking at the evolution of the hominin brain, we will find the selective pressures that led humans to have the type of brain that we have. What proved to be adaptive in our foraging past makes up our biological inheritance, and the development of culture has been built on this heritage. I will argue that the evolutionary pressures faced by our hominin predecessors paved the way for emerging humans to have what can be termed a shared intentionality (Tomasello 2014). From this, I will argue that humans have a 'we-mode' attitude that involves thinking and acting from the perspective of the group (Tuomela 2002).

In Chapter 4, I will argue that the main distinction between humans and other animals is that humans comprehend situations and have the capacity to reason. One of the problems with the BGP narrative is its emphasis on the free-rider problem, presuming that a rational person must single-mindedly pursue their own welfare. Since the BGP take for granted that people have only their self-interest in mind, they assume there is a need for some type of supernatural threat to keep people honest. I argue that people have a host of reasons why they act in certain ways, and often it is with others in mind. Being a rational or reasonable person is being able to subject one's decisions to critical examination (Sen 2009). That is not to say that emotions play no role in guiding our actions.

The work of economist Robert Frank suggests that the role of emotions is to solve what he calls 'commitment problems'. A commitment problem 'arises when it is in a person's interest to make a binding commitment to behave in a way that may later seem contrary to self interest' (Frank 1988, p. 47). For Frank, a moral conscience is a guarantee that a person will follow through with good behaviour even when it may be irrational to do so. Part of being a good potential partner is to demonstrate that you can be relied upon. It follows that those who demonstrate commitments will be judged favourably in interpersonal relationships. While emotions have served us well in making trustworthy commitments, this does not mean that our moral views come from our intuitions. Sometimes people are influenced by what others in the group believe.

Furthermore, I will argue that the evolution of human conscience began with the social control by other people and that this was initially non-moralistic. This led individuals who were better able to internalize the rules of the society more likely to succeed than those who failed to do so. Moreover, since humans have a penchant to dominate others, there would have been pressures to evolve tendencies to resist domination. This type of punitive control would have affected gene pools (Boehm 2012).

While in Aarhus, Denmark, Ara Norenzayan was surprised to find that anyone could simply borrow a bicycle, free of charge, from any number of points of distribution within the city. He was puzzled by the fact that no one stole the bicycles even though most people don't believe in gods, as Denmark is one of the least religious societies on earth (Norenzayan 2013, p. 170). How could this happen since the thesis of the BGP is that belief in punishing Big Gods was the 'driving force behind large cooperative social groups' (Norenzayan 2013, p. 170)? Norenzayan questioned how it was possible for the Danes to keep the wheels of cooperation going without a threatening God? He answers his own question by suggesting that governments with a strong rule of law passed a threshold no longer needing religion to sustain large-scale cooperation. Norenzayan claims, 'secular societies climbed the ladder of religion, and then kicked it away' (Norenzayan 2013, p. 172).

The narrative offered by Norenzayan doesn't support my understanding of the historical record. I will argue that democratic secular societies are fundamentally different from religious societies, and in order to have the type of cooperation that I describe as 'thick cooperation', there must be a recognition of individual rights and human autonomy. This, I will argue, did not begin to take shape until the eighteenth century. While the BGP argue that the belief in God persuades people to cooperate, I will argue that even though people naturally cooperate, in order to achieve 'thick cooperation' there is a need at times to challenge long-standing traditions. Greater cooperation was achieved by identifying the injustices on the ground and offering alternative solutions to make life a little less unjust for more and more people.

The evidence from developmental psychology supports the claims that I make in this book. In Chapter 5, I will provide evidence demonstrating that children naturally help and share with others. Furthermore, much of their prosocial behaviour was not in response to reward or recognition and was not done so to avoid punishment. In fact, there is evidence to suggest that what is important for intrinsic motivation to help or share with others is in having the ability to have a choice (Hepach 2013). What we will find is that children will cooperate more if they are not threatened to do so but rather are provided with the opportunity to make their own choices. Being given the opportunity to make one's own choice is a necessary ingredient for achieving 'thick cooperation'.

I will also argue that researchers have demonstrated a clear distinction between violations of conventional norms and violations of moral principles (Turiel 1983). An example of the former would be a prohibition against texting in class, while the latter concerns rights, justice and the welfare of others. The evidence offered in the fifth chapter suggests that children reflect and reason about issues concerning rights and the welfare of others and have a normative understanding about these issues regardless of what authority figures say. For instance, if God said it was ok to hit another person, the vast majority of children reason that it would still be wrong (Nucci 2001). In contrast, when questions turn to nonmoral judgments, such as worshiping on certain days of the week, these children reason that contravening these conventions would not be wrong, unless God or a religious authority declares it to be so (Nucci 2001). The evidence from developmental psychology suggests that the majority of children naturally cooperate and take offence and intervene when the welfare or liberty of others is at stake. Furthermore, these studies demonstrate quite clearly that children do not need to be threatened to act prosocially.

In the final chapter, I question the approach taken by a number of contemporary cultural evolutionary thinkers who resort to group selection. In particular, I take issue with the claim that 'non-negotiable sacred values' that are grounded metaphysically are what holds societies together and therefore are instrumental for expansive cooperation. In contrast, I suggest that we must be very cautious when talk turns to 'sacred values', and I will offer historical examples of how brave people have fought for the rights of others, leading to 'thicker cooperation' despite the 'sacred values' of the time.

Before looking at the evidence in order to solve the evolutionary puzzle of human cooperation, I will first make clear my use of a few terms. I will use the term gods with a lower case g when I am speaking about supernatural agents. Supernatural agents or gods are invisible forces that are not present in the world of everyday experience but occupy the metaphysical realm that is supposedly beyond this world. The BGP claim rests on the idea that social monitoring is outsourced to the gods as they can watch when no one else is watching and threaten when no one else can (Norenzayan 2013). This allegedly causes people to act prosocially. Ostensibly any god can achieve this, but the BGP hypothesis asserts it is a High God who is omniscient, all powerful and morally concerned with the welfare of humans who is the force for the expansion of cooperation. Because of our mental tools, we have the capacity to believe in any gods, but according to the BGP, the 'God of the Abrahamic traditions may be one of the more well-suited god concepts for cultural survival' (Barrett 2004). Therefore, the use of the upper case G God will be reserved for the belief in one God who is ostensibly concerned with human morality.

The other term needing clarification is the word hominin. For many years, scientists have used the word hominid to refer to all the species that were bipedal and had hands that were free to carry objects such as tools. This terminology changed quite recently as advances in deoxyribonucleic acid (DNA) analysis led researchers to determine the genetic distance between one species and another (Currier 2017). The evidence suggests that about 10 million years ago, with the drying of Africa, many species of apes died, while the number of monkey species increased. One ape lineage managed to survive, and it is the common ancestor of all the living great apes found today (Dunbar 2016). About 6 million years ago, the lineage that ultimately led to humans diverged from the Last Common Ancestor (LCA) of humans and chimpanzees. Approximately 2 million years ago, the genus *Homo* appeared and at the same time the chimpanzee lineage split into the common chimpanzee and the bonobo. Today, scientists refer to the great ape family that includes humans as hominids, while the term hominin is reserved for all the members of the lineage after the LCA. For the purposes of this book, I will use this distinction. With this clarification in mind and since a fashionable solution to the evolutionary puzzle of human cooperation involves a belief in God, I will begin the first chapter by questioning why people would have such a belief.

1

Minds, gods and group selection theory

According to the standard model of the CSR, the human mind evolved modules that were adaptive to the Pleistocene conditions of our forebears, and due to the influence of these mental mechanisms, religion is possible (Lawson and McCauley 1990). That is, religious thought and accompanying behaviour are natural due to the structure of the human cognitive architecture. Yet, there is considerable debate between the different schools within the CSR, particularly when it comes to the conversation about the innateness of religious belief. For instance, Pascal Boyer argues persuasively that religious ideas are parasitic on the naturally evolved processes of the human mind, and although religious ideas are found virtually in every culture, these ideas are not inevitable (Boyer 1994). Like Boyer, Justin Barrett makes the case that religious ideas are the product of ordinary cognitive processes; however, Barrett claims that religion, particularly monotheistic religions, is not only inescapable but also, to a large extent, beneficial for humanity (Barrett 2004). Still others argue that belief in supernatural agents was biologically selected for due to their beneficial effects on cooperation (Bering and Johnson 2005; Bering 2007; Johnson 2016). Jesse Bering goes so far as to suggest that 'God is a way of thinking that has been rendered permanent by natural selection' (Bering 2007, p. 168). Finally, some contend, although not a biological adaptation, belief in supernatural agents or gods is a functional adaptation at the level of the group (Wilson 2002, 2015; Norenzayan 2013).

Despite the differences in the diverse methodologies in the study of religion, and consequently, the multiple interpretations derived from the data produced by the CSR researchers, they all tend to use phrases and concepts that are not entirely clear. For instance, CSR researchers repeatedly use phrases such as 'religious thinking', 'religious dispositions' and 'religious concepts'. While these phrases lead to ambiguous claims, a more fundamental concern is how does one satisfactorily define religion? William James interpreted religion as 'the feelings, acts and experiences of individual men in their solitude, so far as they apprehend themselves to stand in relation to whatever they may consider divine' (James 1902, p. 34). Emile Durkheim, on the other hand, considered religion as 'a unified system of beliefs and practices relative to sacred things, that is things set apart and forbidden' (Durkheim 1912, p. 44). The former emphasizes the so-called mystical experiences of individuals, while the latter stresses the importance of religious doctrines or institutions. Today, the definition of religion used by James is associated with spirituality, while the definition adopted by Durkheim describes the religious institution.

Both these terms, religion and spirituality, are so vague and refer to such a wide variety of phenomena that they are very difficult to define clearly. Traditionally, in the comparative study of religion, scholars presumed that religion was a universal phenomenon that had been interpreted differently by all cultures. It was as if there existed an objective feature of all societies that could be grasped, studied and compared. Even today, some argue, 'for very good scientific reasons, emphasizing cross-religion similarities is the way the science of religion is typically done' (Barrett 2004, p. 75). In contrast, I suggest that it is quite hopeless to try to identify a universal essence of religion. Sam Harris points out that attempting to find the 'unity of religions' is confusing and suggests that the word religions is comparable to the word sports. To suggest that sports is a universal activity does not help us in understanding what the athletes actually do. He suggests that what sports have in common, besides the fact that the participants breathe, is not very much (Harris 2014).

Furthermore, other scholars who studied religion refused to consider it as something to be explained but elevated it to a position that suggested that it was different from any other human behaviour. One influential religious studies scholar argued that 'no statement of religion is valid unless it can be acknowledged by that religion's believers' (Smith 1959, p. 42). So, it was claimed that unless you were a part of the particular religion, you had no chance of understanding what it was really about. The term religion then constituted something so special that it was sacrosanct and indescribable.

The term spirituality is an even more difficult term to unpack. You may hear people say, 'I am not very religious but I am spiritual'. What exactly does this mean? Historically, spirituality is a term that has been associated with values or morality, but seemingly not concerned with material 'stuff' like elaborate homes, fast cars or a fondness for dressing expensively. However, one does not necessarily have to have a belief in a supernatural world to have an awestruck vision of life, grateful for the innumerable wonders of the world and quite content with the vicissitudes of daily life. To be spiritual does not necessarily have anything to do with the supernatural or the gods. I argue that the majority of people want to live a good life, and rather than suggesting that they are spiritual, we should just simply say they are human.

Rather than concentrating on religions, the first part of this chapter aims at an explanation for the belief in gods. The reason for focusing on gods is threefold. First, as just mentioned, there are just too many problems associated with defining religion, and arguably, the idea of god is less ambiguous than the term religion. Second, it appears that most, if not all, cultures throughout history have had a belief in a supernatural world, and within this world, there were agents with 'strategic information' about the states of affairs in the ordinary world (Boyer 1994). Third, and most importantly, ultimately we want to determine why people are better at getting along at the level of civilizations, and the argument from many of the CSR researchers is that prosocial religions with their 'belief in Big Gods who, watch intervene, and demand hard-to-fake loyalty displays, facilitated the rise of cooperation in large groups of anonymous strangers' (Norenzayan 2013, p. 8). Therefore, to determine whether the idea of god or God was a good buy in the marketplace of world views, we need to establish why people could have such convictions in the first place.

This chapter will be outlined in the following manner. First, it will be argued that evidence from cognitive psychology and neuroscience suggests that within the human brain there are two cognitive systems that have considerably different functions. The two processes have been labelled System 1 and System 2. The former involves intuitive processes that serve the goals of the genes, while the latter is viewed as a controlled process that serves the goals of the individual. While our ideas about gods arise from our intuitive thinking, these thoughts are not held in isolation. They are shared narrative accounts of the experiences of encounters with what are believed to be supernatural agents. I will argue that the stories told about such supernatural agents and gods are influential because they are deemed to be important by the particular community. From there, I will look at the argument proposed by some researchers incorporating a so-called group selection approach in the study of religion who claim that groups made up of people who believed in god outcompeted other non-god-fearing groups. Historically, there has been controversy about the idea of group selection in evolution generally, but when it comes to its use in religion, I contend that the evidence provided is simply too weak to be taken seriously. I will begin with the dual-process theory of human cognition.

The prevailing view among many neuroscientists and cognitive psychologists is to conceive the human mind as involving two distinct sets of processes (Lieberman 2007). The automatic system is spontaneous, reflexive and involves emotions and arises early in evolution and development. The second system is controlled and is slower, takes more effort, involves declarative and reflective thinking and arises late in evolution and development. Although there is a general agreement that human cognition involves two distinct processes, there are many ways of describing it. For instance, Matthew Lieberman calls the reflexive or automatic structure the X system, while the reflective or controlled component is referred to as the C system (Lieberman 2007). Others have differentiated the two as heuristic versus analytic (Evans 2010), automatic versus conscious (Bargh and Chartrand 1999) or intuitive versus reasoning (Haidt 2001). More recently, psychologist Joshua Greene equated the dual process with a camera, equipped with both a point and shoot setting and a manual mode (Greene 2013). The automatic option on a camera is not very flexible but is highly efficient and fast, while the manual mode takes longer but gives greater flexibility.

System 1 – The autonomous set of systems

One of the pioneers for developing the dual process of cognitive functioning is Keith Stanovich, and his theory of human cognition is similar to Greene's without the problems associated with the camera metaphor. Stanovich suggests that within the brain, there are two cognitive systems with 'separable goal structures and separate types of mechanisms for implementing the goal structures' (Stanovich 2004, p. 34). He appropriately calls System 1 as The Autonomous Set of Systems (TASS) as this counters any connotations of a single cognitive system. Rather than a single structure, the idea of TASS suggests a number of systems unified by the fact that they are all triggered autonomously by both internal and external stimuli and are not under the control of the analytic processing system. TASS is evolutionarily older than System 2 and was fashioned by solving the problems faced by our ancestors throughout our evolutionary history. TASS is characterized as fast, automatic, heuristic, mostly beyond awareness and effortless and serves the goals of our genes (Stanovich 2004, p. 44–45).

TASS includes encapsulated modules for solving adaptive problems in our evolutionary history, emotional regulation and 'processes of experiential association that have been learned to automaticity' (Stanovich and Toplah 2012, p. 8). Where this theory differs slightly from some evolutionary psychologists is in the claim that although innate modules are an essential component of TASS, they also include processes of behavioural control by the emotions and the fact that some processes are influenced by experience and practice. The important point here is that the property of autonomy can be acquired. That is, problems can be contextualized in the light of one's prior belief and knowledge. The key features of TASS therefore include automatically responding to domain-relevant stimuli, that the execution is free from an analytic processing system and that TASS and System 2 processing are occasionally at odds (Stanovich 2004, p. 37).

TASS autonomously responds to stimuli and primes the organism to be ready. Once the stimuli are detected by one of the subsystems of TASS, they cannot be 'turned off' (Stanovich 2004, p. 52). TASS continuously offers suggestions to System 2 in the form of intuitions and feelings. So it is not only that TASS is generating responses on its own, but it is also providing information from both the social and physical environments in order for the analytic system to make sense of the world. When the two systems are at odds, we typically default to System 1 processing as it takes less computational effort. Humans therefore are, at times, 'cognitive misers' (Stanovich 2011, p. 30).

System 2 – processing

In contrast to the operating system of TASS, System 2 processing is characterized as slow, controlled by a central executive, effortful, under conscious awareness, rule-based and serves the goals of the individual (Stanovich 2004). TASS feeds System 2 with information about the social and physical world, and from that information, System 2 fosters a coherent story. One of the key features of System 2 is the ability for hypothetical reasoning. Hypothetical reasoning allows the individual to postulate different conceivable scenarios about the state of the world, and this acumen is required for such behaviours as deductive reasoning, decision-making and scientific thinking: the mind tools necessary for rational thought.

A good example of System 2 reasoning can be demonstrated by a brain-teaser, sometimes referred to as the Ann problem. Consider the following information. Jack is looking at Ann, but Ann is looking at George. Jack is married, but George is not. The question posed is whether a married person is looking at an unmarried person. One can respond with either yes, no or cannot tell. Over 80 per cent of the people who participated responded by indicating that they could not tell, which is incorrect as the correct answer is yes (Stanovich 2015).

The riddle can be solved by using disjunctive reasoning, that is, reasoning that includes all the possibilities. Since we do not know whether Ann is married or not, but we do know the marital status of the two men, we need to think this through by looking at the question with the possibility that Ann is married *and* with the plausibility that Ann is not married. By using this method of thinking, we surmise if Ann is married, then a married person (Ann) is looking at an unmarried person (George). Alternatively, if Ann is not married, then a married person (Jack) is looking at an unmarried person (Ann). From this, we have determined, yes indeed a married person is looking at an unmarried person. Clearly, this type of reasoning is slow, certainly under conscious awareness, rule-based and demanding in the sense that one must simultaneously hold separate hypotheticals in mind and therefore requires a working memory.

Many of the dual-system accounts divide the two ways of thinking; however, Stanovich takes it one step further and suggests that System 2-type thinking should be distinguished between the algorithmic and reflective mind. He defends this position effectively by pointing out the extensive literature on individual differences between cognitive ability and thinking dispositions (Stanovich 2011). He provides evidence by identifying the mass of cognitive biases that have little to do with cognitive ability (Stanovich 2011). For Stanovich, this is quite evident when studying intelligence tests, as these assessments only capture the algorithmic ability and not the ability of the reflective mind. That is why smart people make common mistakes.

It has been argued that one of the main functions of System 2 is to override TASS. The problem with much of the literature on thinking biases erroneously assumed that System 2 processing failed to override our TASS impulses. Stanovich suggests that if System 2 was never called upon in the first place, then it should not be considered an override problem. To help articulate how to explain override problems in System 2, he uses David Perkins' terminology of mindware. By mindware, he means probabilistic reasoning, scientific reasoning, falsification and causal reasoning. One learns these types of reasoning from experience; however, if one does not have the information or rule, then one would not have the ability to reflect on alternative views. Stanovich concludes that the first default in System 2 is to employ a serial associative cognition with a focal bias. By this he means that the individual fails to fully develop a mental simulation. A good example of serial associative cognition with a focal bias is captured in studies using the Wason Four Card Selection Task (Figure 1).

Participants are shown four cards as illustrated below. They are told that each card has a number on one side and a letter on the reverse. The rule simply

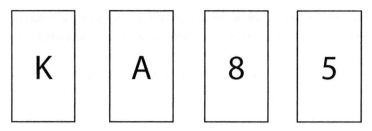

Figure 1 Wason Four Card Selection Task.

states that when a card is shown with a vowel on one side, then there will be an even number on the opposite side. The participants must determine what card(s) must be turned over in order to determine whether the rule is true or false. To arrive at the correct answer, the participant must turn over two cards, A and 5. The majority of participants, however, select A and 8. In interpreting the results from these types of tasks, Jonathan Evans has called attention to the fact that researchers have incorrectly concluded that there is no Type 2 processing involved but only Type 1 processing (Evans 2008).

By analysing the results of these studies, Stanovich found that most of the subjects were engaged in slow, methodical processing. Problems occurred, however, when subjects failed to consider any other possibilities. The majority of subjects did not hypothesize a situation where the rule could be false. Indeed Wason was interested in determining whether people were good at falsifying hypotheses. Stanovich suggests that the problem lies with the fact that participants assume the rule true and then worked through what would happen only if the rule was true. To arrive at their answer, the subjects were not using automatic processing but were nonetheless 'inflexibly locked into an associative mode that takes as its starting point the model of the world, that is the one most easy to construct' (Stanovich 2011, p. 65). The individual begins with a focal model and never considers deviating from it but simply generates associations, failing to generate an alternative hypothesis. Reasoning in this way leads people to 'represent only one state of affairs' (Stanovich 2011, p. 66). Instead of considering all possible alternatives, subjects in the Wason Four Card Selection Task start with the given rule and then attempt to confirm their assumption.

How this assists us with understanding the persistence of belief in gods is the fact that TASS processing gives us the potential to conceive of gods and other supernatural agents, and our algorithmic mind creates and circulates the stories told about these beliefs. However, a reflective mind involves falsification and scientific reasoning, mind tools that are not always used to assess the rationality of the stories about the gods. I will argue that in addition to having a reflective mind, a necessary condition for reflective thinking is an environment where this type of reasoning is permitted and, more importantly, encouraged. Later in this chapter, I will argue that we find evidence for the necessary environment for free thought and reflective thinking with the Milesians, a group of freethinkers from the sixth century BCE. First, however, the relationship between TASS and God concepts will be discussed presently.

TASS and God concepts

CSR researchers have identified a number of TASS system modules to account for the ubiquitous belief in supernatural agents or gods. They suggest that much of our thinking about gods arises from interpreting the world through TASS. Justin Barrett doesn't use the same wording, but he clearly separates beliefs into reflective and non-reflective, with the former being held consciously arising through deliberate reflection, while the latter comes automatically and seems to arise instantaneously. Barrett suggests that 'our minds produce non-reflective beliefs automatically, all the time' (Barrett 2004, p. 181). What influences our beliefs, including what we think exists in the world, arises from mental tools that dwell below our conscious radar (Boyer 2001, p. 107). For Boyer, 'when creating what might be called reflective beliefs, unless given strong reason to the contrary, we simply adopt these non-reflective beliefs as reflective beliefs' (Boyer 2001, p. 107).

The reason for the belief in gods, suggested by these theorists, is simply due to the fact that such belief is supported by our intuitive mind tools. How we come to think about the characteristics of the gods attractively fit with what people understand about the nature of the world. People are outfitted with innate intuitive theories about how the world works. In fact, it seems that how humans acquire knowledge about anything is influenced by our predisposed intuitions in the separate domains of physics, biology and psychology. For instance, within the physical domain, children have intuitive knowledge about the characteristics of objects. They have a sense of solidity, gravity and movement. Research suggests that children at a very early age would be surprised to find objects jumping over or passing through other objects or surfaces (Spelke 1994). Additionally, children intuitively believe that objects can only act upon another object if they come into contact with each other (Spelke 1994). Finally, children who are put in front of a flight simulator and tasked to bomb a target on the ground tend to miss the mark. Intuitively, children wait until they are over the target, failing to consider that when the bomb is dropped, it will continue forward at the pace of the aircraft, drastically affecting where it will land. This has also been demonstrated in experiments with university students (McCloskey et al. 1980). So, according to CSR researchers, just as it is natural to have intuitive beliefs about the world, the idea of god comes just as naturally.

No organism, including humans, could possibly consider every object in the world as something unique. Therefore we place objects into categories or what Boyer calls templates (Boyer 2001). In the biological domain, evidence suggests that people have a clear distinction between animate and inanimate objects (Pinker 1997). People have intuitions about animals and plants that are different from intuitions about artefacts. For instance, infants behave as if only animate beings are capable of self-propelled motion. Animals and people have agency, and these agents are thought to have an 'internal and renewable source of energy, force, impetus, or oomph, which they use to propel themselves, usually in service of a goal' (Pinker 1997, p. 322). Furthermore, children believe that the different types of animate beings share similar properties and 'living kinds activates an essentialist principle' (Boyer 2001). For instance, children still consider a horse a horse even if it is dressed in zebra pyjamas (Keil 1994). They believe that there is some kind of hidden trait or essence that all horses share. Scott Atran provides evidence that this classification of the natural world is universally shared by everyone (Atran 1990).

Since humans are social animals, those who were better at predicting the behaviour of others would have an adaptive advantage in the social environment of our forebears. Evolutionary psychologists argue convincingly that humans developed an intuitive psychology to account for interactions with others (Pinker 1997). Humans have what is referred to as a Theory of Mind (ToM) or folk psychology, in that we have the ability to make automatic, unconscious inferences about the beliefs, desires, motivations and intentions of others (Gopnik 1999). Rather than using ToM, Daniel Dennett prefers the term intentional stance to account for such behaviour (Dennett 2013). The reason for this distinction is that the word theory implies hypothesis testing, and this is not what the idea really means. An intentional stance is 'about what agents are thinking about. It is not something that we intentionally adopt in the normal course of our day, it is typically involuntary and seldom noticed' (Dennett 2013, p. 78). For Dennett, the intentional stance is something that we do not have to teach our children; rather, it comes naturally.

Conceiving the world as home to numerous agents contributes to another important tool that operates automatically, namely, a Hypersensitive Agent Detection Device (HADD) (Barrett 2004). That is, people tend to interpret ambiguous evidence as being caused by an agent. The evidence suggests that humans are remarkably sensitive to cues of animacy. This is part of TASS, and from an evolutionary perspective, it makes sense. Whenever we are startled by a noise, we anticipate a possible predator or agent, rather than the wind or some other innocuous source. Those who are prone to look for agents in the world are more likely to survive and pass their genes on to the next generation.

The reasoning follows that the belief in gods stems naturally from the functioning of these completely normal mental tools operating in social contexts. Barrett argues that it is reasonable to understand why people believe in gods by simply taking the template for person and adding to it 'counterintuitive properties', such as the ability to walk through walls (Barrett 2004, p. 21). These counterintuitive properties are attention grabbing, and when combined with a tacit inference, there is evidence to suggest that they may be easier to remember (Boyer 1994). This combination is culturally widespread as well (Boyer 1998). Moreover, in addition to the attention-grabbing counterintuitive properties, gods or supernatural agents hold 'strategic information', that is, their omnipotence and omnipresence allow them to know salient information about the natural world (Boyer 2001, p. 107). There are a number of studies providing evidence that the idea of god relies heavily on our folk psychology (Barrett 2004; Atran 2002). Since our HADD makes us prone to look for agency in the environment and thus we naturally attribute this agency to gods, some studies suggest that humans are equipped with a supersense (Hood 2012).

Supersense

Psychologist Bruce Hood argues that humans are in fact a sacred species. He points out that we have brains that automatically perceive faces and structures in the world, which naturally leads us to believe in the supernatural (Hood 2012). Indeed, researchers have identified a cortical region of the brain, called the fusiform face area, that is activated when people look at faces (Kanwisher 1997). Once triggered, our domain-specific face recognition module cannot be turned off. For Hood, however, the spreading of supernatural beliefs cannot be pinned solely on religion or culture, as ideas about god simply capitalize on our supersense. What religion and culture provide, according to Hood, is the necessary framework for what comes naturally to us.

Hood draws on evidence from developmental psychology to support his claim that our brains are designed to infer patterns in the world and that we make sense of them by generating intuitive theories. For instance, he asks whether you have ever had the experience of feeling as though someone was watching you. He suggests that everyone has had these experiences, and he would be surprised if someone had answered in the negative. He says that this ability would have great adaptive value, but in fact, we don't really have this ability. Although it is not a true human ability, Hood insists that we have this 'sense'. The sense of being stared at, according to Hood, is an example of our supersense. It is important to stress here that the supersense is actually a minor misconception or a false belief. For instance, sometimes you may happen to look around and see that someone is staring at you. You deduce from this that you must have had a sense of being watched, otherwise why would your attention be drawn to the person in the first place? The most feasible explanation for why this occasionally occurs is simply due to the fact that after you catch someone looking at you, you may infer that you had a sense of being watched. In actual fact, you can stare at someone as long as you like without being noticed.

Psychologist Daniel Simons, co-author of *The Invisible Gorilla,* demonstrates this intuition very well by conducting a clever experiment. Simons wanted to understand what people intuitively believe about how the mind works. He stopped people on the street and asked them whether they strongly agree, mostly agree, mostly disagree or strongly disagree to a number of questions. One of the questions was whether they believed that a person was capable of sensing of being looked at. The majority of people strongly agreed with this statement, and while they were answering the question in the affirmative, they failed to sense a person dressed in a gorilla suit, immediately behind them, who was staring at them. Of course, you can tell people that no one has the ability to sense being looked at, but some will insist that humans have the capacity to sense being watched.

Similarly, some people believe events happen in threes. In reality, events happen in fours, fives or sixes depending on where you draw the boundary. If people believe events happen in threes, they will inevitably find this pattern. For instance, if a person hears that two people have recently passed away and the person is convinced that events happen in threes, then any death that the person hears about will confirm their belief. It doesn't matter how well one knows the person who passed away; it could be an actor, a politician or a friend of a friend of a friend. They will still take this as 'evidence' that supports their hypothesis. Humans are, at times, victims of illusory correlations by seeing intentional causation in random processes, and this is important to understand. However, more importantly, we need not cherish the cognitive processes that lead to these errors, as it could lead to problematic assumptions.

For instance, Hood concludes his book by suggesting that although we have the capacity for both TASS and Type 2 processing, intuitive reasoning has an advantage as it is effortless and fast (Hood 2009). He says that it has served us well in our evolutionary history, and the evidence for its value is the fact that we are here speaking about it, so we should learn to foster it. He adds that our supersense will not disappear in the modern period as 'it makes possible our commitment to the idea that there are sacred values in our world' (Hood 2012, p. 250). While there is no doubt that TASS has served us well in our evolutionary past, there are a number of problems with the conclusion that we should foster our Type 1 thinking and cherish the so-called sacred values.

Emotional dogs and rational tails

Hood joins company with a growing number of researchers who claim that there is far too much emphasis placed on reasoning rather than intuition when assessing how we reach our moral decisions. For instance, Jonathan Haidt argues, in his often cited 'The Emotional Dog and Its Rational Tail', that reasoning is rarely the cause of our moral judgments and that our decisions are emotionally driven (Haidt 2001). For Haidt, our reasoning takes place after our decisions are made, and we typically use reasons to justify our actions. Haidt not only claims the idea that people make moral judgments through a process of reasoning and reflection a Western delusion, but also insists that 'anyone who values truth should stop worshipping reason' (Haidt 2012, p. 104). His theory suggests that there are moral truths that are discovered through a process that is more akin to perception, in which 'one just sees, without argument, that they are and must be true' (Haidt 2001, p. 814). For Haidt, moral reasoning, on the other hand, is an *ex post facto* process that is essentially used to influence others. In this model, one feels an immediate emotional reaction such as revulsion and then '*knows* intuitively that something is wrong' (Haidt 2001, p. 814 *my italics*). Apparently, 'one sees or hears about a social event and one instantly feels approval or disapproval' (Haidt 2001, p. 818). For Haidt, when it comes to moral judgement, it is the emotional tail that wags the rational dog (Haidt 2001).

Although Haidt is suggesting that our emotions play a greater role than reasoning, he does admit that this is 'offered as a plausible alternative approach to moral psychology, not as an established fact' (Haidt 2001, p. 815). More importantly, he says that despite the fact that moral reasoning is rarely the direct cause of moral judgement, his model is a 'descriptive claim about how moral judgements are actually made, it is not a normative claim offering how moral judgements ought to be made' (Haidt 2001, p. 815). Despite the cautionary tone, many researchers maintain that our intuitions about morals must be given their due expression and that culture affords the context for this expression (Csikszentmihalyi 1990; Slingerland 2014).

Haidt's social intuitionist model differs fundamentally from others who suggest that we arrive at our moral decisions through a process of reasoning and reflection (Piaget 1932/1997; Turiel 1983). For instance, psychologist Elliot Turiel makes a compelling case against the intuitionist position, suggesting that much of the evidence to support the claim that moral judgements arise from Type 1 thinking do not really involve moral issues (Turiel 1983). The social domain theorists make a clear distinction between social conventions and moral issues. Social conventions are defined as existing social arrangements that 'receive their legitimacy either through standards agreed upon by a community or through its established systems of rules, customs or authority' (Helwig and Turiel 2017, p. 138). In contrast, making judgements about moral issues is not a reflection of either predetermined biological dispositions or the acceptance of cultural norms but rather involves issues related to justice, rights and the welfare of others. There is a wealth of studies suggesting judgements about such acts as inflicting harm on others involve complex discernment (Nucci 2001). Furthermore, much of the evidence offered by Haidt comes from responses by university students faced with bizarre situations that are not true to everyday-life circumstances. Some of the studies involve incest, masturbating with chickens, cannibalism and eating food shaped like dog turd (Haidt 2012). In contrast, the social domain theorist research includes situations that involve acts of unfairness, discrimination, inflicting harm and social exclusion (Nucci 2001; Killen and Rutland 2013). These types of situations are the types of events that happen on a daily basis. Larry Nucci and others have found in these types of situations that complex reasoning is definitely involved; however, in particular situations, the responses may be quick (Nucci 2001).

Psychologist John Kihlstrom aptly calls the approach by such theorists as Haidt as the people-are-stupid-school of psychology, as they advocate that people simply rely on their biases and heuristics without thinking very much about anything (Kihlstrom 2004). According to Kihlstrom, this school claims that people are non-rational creatures, relying on emotion and come to decisions mostly without reflection. Since our decisions arise out of the processes of Type 1 thinking, they are non-conscious in nature and do not involve making a choice (Kihlstrom 2004). Much of the evidence to support their argument comes from priming experiments, and I will have much more to say about the problems with this type of methodology in the following chapter.

By neglecting the evidence provided by the social domain theorists that suggests people reason much more about moral situations than Haidt gives them credit for, psychologists such as Barrett and Hood jump to the conclusion that we should get our values from our intuitive psychology. Why does it follow that we should get our values from what Hood labels a supersense? Barrett makes a similar claim when he suggests theists have it easy, since they can casually make moral judgements, and while, at times, they may feel upset at having been wronged, they can be rest assured that God will reward the just and smite the wicked (Barrett 2004). For atheists, on the other hand, Barrett counters that things are much more difficult. The atheist has the same intuitive sense of morality as the theist does but does not have the moral certitude (Barrett 2004, p. 110). In Chapters 4 and 6, I will offer evidence to demonstrate the problems associated with moral certitude. Nevertheless, Barrett goes on to suggest that the atheist has to 'concoct theories of morality' that justify their moral certainty (Barrett 2004, p. 110). Barrett doesn't specify these contrivances, but if he means such things as individual rights, and freedom of speech, then I would argue that these are very good theories to concoct. Indeed, a major claim in this book is that human cooperation can only reach the level of 'thick cooperation' from the features of society that humans devised in spite of our passions and feelings. Rather than advocating a moral certitude, what ultimately leads to a just society and 'thick cooperation' is one that offers an environment that invites a challenge to certitude. I argue that problems arise with moral definitiveness.

Turiel offers a compelling historical example to show the weakness in advocating moral certitude and relying on Type 1 thinking. Turiel asks us to take our minds back to a time between 1920 and 1950 in a southern state in the United States and consider an African American man and a Caucasian woman deciding to make love (Turiel 2015). Turiel accurately suggests that many from the white population would have had a powerful, immediate sense that this act was wrong. I would add that they probably were absolutely certain that it was wrong. They would have 'felt' that white and black people should not be romantically involved. They would have had a gut-reaction to this assault on tradition. For most people, in that particular time period, these *moral truths* were self-evident. Just imagine if the two characters in Turiel's example were both male or both female and consider how people would have 'felt' in that type of situation. Clearly, from this example, one can see that there is much more to our judgements than intuitions. Furthermore, examples like this suggest that when making claims about states of affairs in the world, without reasoning and reflection, situations could easily get ugly.

It has been argued that humans have the type of mind that makes the idea of gods or supernatural agents plausible. While some scholars, such as Barrett, attribute such belief to personal experience, I argue that people don't come to believe in gods simply due to the false positives produced by HADD and the ability to theorize about other minds. After all, humans have the type of mind to speculate about green martians with strategic information as well, but these types of ideas are not important unless they are deemed to be important by other people. Rather, people are influenced by stories involving supernatural agents, especially when the stories come from the most influential and powerful people in society. Stories about the gods are not formulated by thinking individuals in isolation; rather, they are brought into existence once people take the opportunity to 'compare notes'.

Shamanism

In attempting to explain why people believe in gods, some have suggested that it has its roots in shamanism. Now, shamanism is another foggy term, but essentially it refers to individuals who have the ability to connect with supernatural agents in an Altered State of Consciousness (ASC) (McClenon 2001). Mircea Eliade suggested that shamanism was universal and that shamans were the social leaders who had the ability to engage the spirit world for healing and other benefits. The essential ingredient of shamanism was the use of ASC to enter the spirit world (Eliade 1964). Others argue that true shamanism should be limited to northern Eurasia and the office of shaman cannot be simply transferred to other cultures (Pentikainen 1996). However, Hays-Gilpin convincingly argues that there is a wealth of evidence to support the claim that shamanism was a common practice in foraging societies and that the shaman, through an ASC, had the ability to have contact with a spirit world (Hays-Gilpin 2004). For Hays-Gilpin, the word shamanism is helpful for both Siberian shamanism and a generalized account of shamanism as a common religious practice in foraging societies, including the concepts of 'layered universe' and a personal contact with the spirit world through an ASC (Hays-Gilpin 2004). Furthermore, anthropologist Michael Winkelman found that the shaman was associated with healing in all the societies he investigated (Winkelman 1990).

The original Tungusic word for shaman comes from the root sa, which means 'to know'. Typically, shamans are specialists who have knowledge about the spiritual world and act as mediators or negotiators between the natural and supernatural realms. Regardless of whether one identifies them as shamans, there is substantial evidence to suggest that some people in the community were believed to have the ability to appropriate the experiences of altered states of consciousness and string it in to a good tale that influenced others (Winkelman 1990, 2011). These people were considered to have privileged access to important information by having the ability to foretell the future, influence the climate and heal the sick. Cognitive archaeologist David Lewis-Williams provides a plausible case for a form of shamanism in the Palaeolithic that may account for the roots of religion.

David Lewis-Williams argues that 'religion sits on the shoulders of a neurological substrate and the experiences that derive from it' (Lewis-Williams 2010, p. 120). He convincingly contends that consciousness should be considered 'not as a state, but as a continuum, from shifting wakefulness to sleep' (Lewis-Williams 2002, p. 126). This spectrum of consciousness includes an ASC that results in different kinds of imagery and experience for each individual. All people have these images, and depending on one's social context, these images can be interpreted differently. The important point being that the forms are 'universal, though their selection and meaning are culturally determined' (Lewis-Williams 2010, p. 146).

Furthermore, in order to sustain the proper functioning of the cells, all mammals and birds need sleep in order for the brain to manufacture proteins. Ostensibly the brain does this at a faster rate when in deep sleep (Lewis-Williams 2010, p.190). When in deep sleep, the brain does not totally shut down, although some parts are 'turned off', such as the dorsolateral prefrontal cortex. This part of the brain is essential for executive functions such as planning and organizing our thoughts. Others have suggested that dopamine is secreted more in rapid eye movement (REM) sleep and therefore our dreams are more emotionally intense (Panksepp and Biven 2012). From this, researchers have speculated that the emotionality of dreams with its hallucinatory character may help us deal with the emotional challenges we face (Panksepp and Biven 2012, p. 140). Nevertheless, evidence suggests that dreams are often bizarre, and while all mammals dream, only humans have the capacity to remember their dreams and therefore the ability to communicate with others about the content of them. Furthermore, when sleeping, the body remains in one position, yet one's dreams can be quite vivid and give the sense that one has travelled somewhere. The idea of a soul may have developed from discussions about dreams and/or death.

Neurologist Kevin Nelson, who researches people who have endured near-death experiences (NDEs), suggests that if we want to understand what people come to term a 'spiritual' experience, then we should look at it from a neurological point of view and determine a physiological account for the tunnel vision and out-of-body experiences. He identifies three states of consciousness: wakefulness, REM sleep and non-REM sleep. Nelson suggests that NDE takes place in the 'borderland' of these three states of consciousness, and perhaps this could also account for the feeling of a 'spiritual' experience. He has found that many of the people who experienced an NDE say that it involved intense light. Ironically when in the REM state, the visual system is strongly activated. Nelson also points out that many people undergoing NDE also experience what he terms REM intrusion. This means that some aspect of REM sleep intruded in their normal day. He goes on to suggest that up to 25 per cent of people experience this phenomenon (Nelson 2011, p. 201). Nelson's research suggests that some people are more prone to experiencing a blending of consciousness and are susceptible to these types of experiences. It would seem that, just as today, some people are better at telling stories and providing meaningful interpretations of dreams and their 'spiritual' experiences. Arguably, the better one was at creating these stories, the more influential the storyteller would be.

Lewis-Williams suggests that the cave art of the Upper Palaeolithic Western Europe provides evidence for, perhaps, the earliest form of shamanism. He suggests that people entered these caves to recreate the experiences of their visions, dreams and hallucinations. He reasons that *Homo sapiens* in the Palaeolithic had brains that were fully modern and therefore had the ability to dream and to have visions. Second, he suggests that the people would invariably try to reach some consensus as to what they mean. He suggests that the caves were used as a tool for social discrimination, giving power to the shamans who had the unique ability to connect with the spirit world. The conclusions that Lewis-Williams draws from the evidence of the cave paintings are interesting. For instance, he points to the caves at Rouffignac to make the case that the image of a horse found there is not just a simple painting on the cave wall but is depicted in such a fashion that the rear of the horse is in the wall, while the head seems to be emerging from the rock.

He suggests that the horse is emerging from the netherworld. This suggests that the wall of the cave could be conceived as a demarcation between the material and spiritual worlds. For Lewis-Williams, the art produced in these caves was 'a complex ritual performance during which image makers engaged, controlled and embodied their visions' (Lewis-Williams 2010, p. 225).

Furthermore, Lewis-Williams makes a plausible case for social stratification by comparing what is found in the Enlene cave with Les Trois Freres, a cave that leads off of Enlene. He suggests that there is evidence that some members of the community had a greater relationship with the spirit world. Those who had access to the deeper areas of the cave were free to touch the sides and perhaps they had 'more intense mental states' (Lewis-Williams 2010, p. 231). Other caves, such as those at Gabillou and Lascaux, appear to have separate sections as well, suggesting that only the elite could gain access. Those shamans who had the natural gift to understand the spiritual world were those who were rewarded with an elevated position in society.

Once some members of the community identified the shaman as one who is worthy of learning from, others would infer that the shaman holds strategic information and would join in to hear the tales. Once the shaman 'cures' someone or makes a good guess on an approaching storm, or receives favourable information from an animal or ancestor spirit, his prestige would increase. There is evidence to suggest that the prestige of a person increases the popularity of the individual and consequently the belief in what the person has to say (Henrich 2016). Just consider how influential some movie stars are on topics that they have very little knowledge about. Some politicians are also given messianic status. Regardless of what you may think of Barack Obama, you have to admit that his deliverance of speeches is second to none. Matt Ridley points out that Obama was so skilful in elocution that he was handed the Nobel Peace Prize essentially for being elected (Ridley 2010, p. 219). The argument here is that in all of human society, there have been some who were very good at telling convincing myths.

Moreover, there is evidence to suggest that prestigious individuals are copied by other individuals (Henrich and Gil-White 2001). People tend to rank others as worthy of copying if they are popular or have a high standing in the opinion of the community. Studies indicate that individuals are influenced by the actions of skilled performers when making investment decisions. For instance, in one study researchers found that when results were posted between rounds, students copied those who were identified as the top performers (Kroll and Levy 1992). Other studies suggest that the stories told by prestigious individuals are remembered longer than the stories told by less-esteemed people. For instance, researchers found that subjects did a better job at retaining information from conversations that involved the 'boss' over conversations held by people with a lower status (Holtgraves 1989).

Additionally, in the courts of law in contemporary society, studies suggest that jurors judge those witnesses who are the most confident as more accurate or credible (Pakosh 2016). While those individuals who express their beliefs most confidently may appear to be credible, there is a substantial body of research suggesting that confidence is only moderately related to accuracy (Cutler et al. 1995). Other studies suggest that there is a strong relationship between confidence and persuasion (Jiang et al. 2000). There are a number of criminal cases where not only were the witnesses mistaken but they were mistaken despite their confident testimony. Consider the case of Anthony Hanemaayer, who was identified by the victim's mother as the person responsible for raping her daughter. She was so convincing that Hanemaayer himself said that if he had been a juror member, he would have found himself guilty (Pakosh 2016). Hanemaayer was eventually exonerated.

The point being made here is not that the shamans were making up the stories for no reason at all. Perhaps, like mistaken eyewitnesses, they believed in the stories they were telling and were extremely confident and successful at persuading others of the 'truthfulness' of their claims. What we find, however, is that the myths told by the shamans were certainly attention grabbing, and since they were told by the most prestigious members in the community, it made the recollection of the myths even more likely. Once the stories of a supernatural world are seen to bring benefits and circumvent problems for members of the society and once they are reinforced by the most influential members of the group, the practices accompanying those beliefs become the norm. And once this takes place, then it seems to be a 'no-brainer' that this is the only correct way of behaving.

It has been argued that all humans have a spectrum of consciousness as elaborated by Lewis-Williams and, furthermore, we all dream. Because of our algorithmic mind, we have the ability to formulate and share stories about these experiences. It is the result of our hypothetical thinking that we have the ability to make and share stories about the gods. For most of our evolutionary history, all humans believed these myths, and they naturally created a narrative account of the world. If it rained, it was due to a supernatural agent, and if one needed rain, then one would pray or dance for it. There existed no alternative method to account for the state of affairs in the world. Don Wiebe argued persuasively that this mythopoeic way of thinking is radically different from scientific or philosophical thought that we are accustomed to today.

Dichotomy thesis

Don Wiebe has argued that there is a distinction between the content and structure of archaic and modern thought (Wiebe 1991). Traditional cultures, he maintains, have an agentic conception of the world and fail to discriminate between appearance and reality. In contrast, the core characteristics of modern thought are critical reflection and analysis. Wiebe argues that Levy-Bruhl's dichotomy thesis brings 'fully to light the mystical aspect of primitive mentality in contrast with the rational aspect of the mentality in our societies' (Wiebe 1991, p. 58). Although his aim was to demonstrate the mythical nature of traditional thought, Levy-Bruhl was by no means suggesting that the modern mind is free from such thinking, as he suggests the difference lies in the fact that it is 'more marked and more easily observable among primitive people than in our own societies' (Wiebe 1991, p. 63). Wiebe argues that we first see evidence of a marked departure from traditional thought and the emergence of modern thinking with a group of theorists in ancient Greece. It is with the Milesians where we find an environment freely inviting reasoned accounts of the world as opposed to narrative accounts.

Emergence of scientific thought

Karl Popper convincingly dates the emergence of philosophical or scientific thought to the sixth century BCE, with a radical group of thinkers living in Miletus, a small town in Ionia (modern-day Turkey). The Milesians, according to Popper, created a tradition of critical discussion that made possible an expansion of knowledge. By examining the slim surviving evidence that we have of Thales, Anaximander and Anaximenes, Popper argues that these thinkers used observations and hypothetical thinking in order to criticize theories. For Popper, what is relevant for epistemology is the study of scientific problems with an invitation for critical arguments and the use of evidence for supporting or refuting those arguments.

A brief look at the theories proposed by the Milesians supports the argument made by Popper. Thales postulated that the primary and ultimate element of the world was water. He reasoned that water is essential for life and that it is the only substance on earth that can exist in all three states (Lederman 2006, p. 35). By abandoning the Homeric conception of nature subject to the wills of the gods and attempting to explain the world by using physical natural laws, Thales paved the way for rational discussion. Thales' thought 'constitutes the first attempt to explain the variety of nature as the modification of something in nature' (Wiebe 1991, p. 91). Of course, Thales was incorrect, but the point here is that he reasoned hypothetically by creating different models of the world. This way of thinking laid the conceptional foundation upon which his successors were able to build.

Thales' student Anaximander countered that the building blocks of matter consisted of what he termed *apeiron*, which means without boundaries. For Anaximander, the earth could not be made simply from water, as the ultimate building block of matter must be able to create, and not destroy, and water according to Anaximander extinguishes rather than generates. Popper says this is 'one of the boldest, most revolutionary and most portentous ideas in the whole history of human thought' (Popper 1968, p. 63). This was so important because Anaximander was free to criticize Thales. In fact, in this tradition, we find that each generation arrives at a radically different conclusion about the elementary units of matter. For instance, we saw that Thales believed the primary substance to be water, Anaximander countered that it was the 'boundless' and a third Milesian, Anaximenes, took the primordial element to be air (Lederman 2006, p. 39).

There seems to be no reason to believe that these philosophers suffered any repercussions due to a departure in thought from their predecessor. What is particularly important here is the freedom to hold differing beliefs and world views. The theories proposed by these thinkers were open to modification and, consequently, we see the emergence of a tradition of critical thought and unbounded by ultimate truths. Popper correctly notes that 'our knowledge and our doctrines are conjectures; consisting of guesses of hypothesis rather than of final and certain truths and that criticism and critical discussion are our means of getting nearer to the truth' (Popper 1968, p. 151).

Wiebe clearly demonstrates that the Milesians shifted their attention away from a narrative account of the world that was guided by the gods to investigating how the world actually worked. They were not concerned with value or trying to derive meaning but were interested in knowledge for the sake of knowledge alone. After the Milesians, we have a 'new way of thinking, a way that is non-agentic in orientation and hence primarily causally oriented and naturalistic' (Wiebe 1991, p. 113). The Milesians 'in some sense constitutes the creation of a scientific world-view that amounts to a disenchantment of the mythopoeic or at least quasi-mythopoeic, world it succeeds' (Wiebe 1991, p. 8). Second, the Milesians were drawing conclusions that seemed to counter intuition. Their explanations were not necessarily derived strictly from empirical observation, but what is truly important is that their analysis had explanatory power. Third, their defence of the propositions required a rational argument, and their reasoning had a cumulative aspect to it. With the Milesians, we find the ingredients necessary for scientific thought.

In the next chapter, I will assess the evidence that purportedly supports the claim that it is the idea of god and a narrative way of thinking that led the way to a larger cohesive society. The argument by some scholars is that a belief in god caused individuals to refrain from antisocial behaviour and therefore was advantageous to both the individual and society. There is an addendum to the argument that the idea of god is beneficial for the individual, and it derives from group selection theory, which states that people in groups, who exhibit traits favourable to the group, outcompete those groups comprising people with less-favourable traits. A leading proponent of this theory is David Sloan Wilson, who has argued that religion was a cultural adaptation as it allowed religious groups to expand and outcompete other social groups (Wilson 2002). In Chapter 3, I will argue that due to environmental pressures, the hominin brain was selected to achieve the advantages of living in groups, but this does not mean that the brain evolved by group selection. I suggest that group selection is unnecessary to account for favourable traits that result in group coherence and especially so when it comes to a discussion about religion. Therefore, I will spend much of the remaining part of this chapter arguing that it is genes and not groups that are the selectional units of evolution.

Cui Bono

In criminal investigations, the police must consider all suspects bearing a motive for the crime. By considering who benefits from the misconduct, the investigator may learn that the person responsible is not the person originally believed to be the primary suspect. The aphorism *Cui Bono* is not only beneficial for the investigator, but also useful for clearing up misconceptions concerning evolutionary theory.

As discussed earlier in this chapter, the older evolutionary structu[re] make TASS slave to the genes, while the architecture of the ana[l] serves the goals of the individual. Stanovich used the term *robot* to illustrate that System 2 serves the goals of the individual, which is analogous to Richard Dawkin's use of survival machines (Dawkins 1976, p. 19). Dawkins decisively argued that humans, as well as plants, bacteria and viruses are all survival machines, created by selfish genes. He suggests that billions of years ago, a stable pattern of atoms formed with the ability to replicate themselves. Dawkins calls these units replicators, and just like any copying process, mistakes happen and subsequently a population of slightly different replicating molecules was assembled. Dawkins suggests that these early replicators 'may have been able to protect themselves either chemically, or by building a physical wall of protein around themselves' (Dawkins 1976, p. 14). To ensure that these molecules would succeed in making copies of themselves, they began to make protective containers. Dawkins uses the term vehicle to label these protective containers, and he defines them as 'an integrated and coherent instrument of replicator preservation' (Dawkins 1982, p. 114). According to this thesis, these vehicles are survival machines. Dawkins convincingly argues that these replicators 'swarm in huge colonies, safe inside gigantic lumbering robots, sealed off from the outside world, communicating with it by tortuous, indirect routes, manipulating it by remote control. These replicators have come a long way and they go by the name of genes and we are their survival machines' (Dawkins 1976, p. 16). This distinction between replicators and vehicles is an essential discrimination when considering the 'success' of adaptations, so it is necessary to make this point clear. A vehicle is anything that houses the replicators, and the success of the replicator is simply to make copies of itself. In contrast, the success of the vehicle is measured by its capacity to house the replicators inside it, in order that the replicators will have the ability to make true copies of itself. Of course, if the vehicle is killed, then everything inside it will die as well, so natural selection, to a certain degree, will favour those replicators that give rise to vehicles that are most likely to survive.

When contemplating the role of any genes in the genome, it is common, even among biologists, to ask what the gene is for. Dawkins doesn't dispute the fact that to understand adaptive behaviour, biologists have typically used 'a gene for x' as a suitable way of establishing a genetic basis for x. Yet Dawkins emphasizes the point that the gene is not for the body but for the gene itself. For Dawkins, the only way to comprehend all the organisms on earth is to conceive them as temporary vehicles being used by selfish genes whose sole purpose is to replicate themselves. While Dawkins considers the body as serving the needs of genes, he is not suggesting that the organism is destined to be selfish. Indeed, he insists, 'genes do not control behaviour directly in the sense of interfering in its performance, they only control behaviour in the sense of programming the machine in advance of performance' (Dawkins 1982, p. 16). If the organism was the replicator, then the organism would be selfish, but this is not necessarily the case. This is a point that is often misconstrued.

For Matt Ridley, one paragraph in *The Selfish Gene* needs to be emphasized. Dawkins wrote, 'it appears that the amount of DNA in organisms is more than is strictly necessary for building them: a large fraction of the DNA is never translated into protein' (Ridley 2010, p. 66). Dawkins, as Ridley points out, goes on to suggest that 'if the purpose of DNA is to supervise the building of bodies it is surprising to find a large quantity of DNA which does no such thing' (Ridley 2010, p. 66). For Dawkins, the simplest way to account for this Junk DNA is to suggest that it is a parasite that freerides in the survival machine, created by other DNA. Although there is controversy over the idea of Junk DNA, Ridley says it is impossible to account for the large proportion of the human genome devoted to Junk DNA and pseudogenes without attributing it to the selfish gene theory. He notes that the most common gene in the human genome is the recipe for reverse transcriptase, which is an enzyme that the human body has little need for. Ridley concludes that Junk DNA provides convincing evidence that the human genome is built by selfish DNA sequences. 'The body is an emergent phenomenon consequent upon the competitive survival of DNA sequences, and a means by which the genome perpetuates itself.' (Ridley 2010, p. 72).

Despite the fact that Dawkins and Stanovich suggest that we are complicated survival machines at the mercy of replicators, they also suggest that we have the capacity to reach beyond the selfish interests of our genes by using the 'mind tools' of reason and science. These tools will allow humans to reflect on their lives and respond accordingly in a way that is unique in the animal kingdom – by 'rational self determination' (Stanovich 2004, p. 9). In contrast to the school of psychologists who suggest that most of our decisions are made without reflection, humans certainly do have the ability to defeat System 1 processing when System 2 processing gives them reasons to do so. This capacity allowed humans and only humans the potential for 'thick cooperation'. I will pursue this logic further in Chapter 4; however, now let us look at another alleged selection process, cultural group selection.

Cultural group selection

Some psychologists and cognitive scientists have argued that a belief in gods that includes reference to both genetic and cultural evolution has greatly supported the expansion of prosocial behaviour beyond mere kinship selection and reciprocal altruism. As previously discussed in this chapter, Type 1 thinking laid the foundation for a belief in supernatural agents, which in turn, according to the BGP, encouraged people to refrain from antisocial behaviour. Some argue that this is a genetic adaptation, as *all* humans innately fear supernatural agents (Johnson 2016). Others suggest that a belief in supernatural agents is a by-product of Type 1 thinking and groups may choose to adopt these belief systems. The groups adopting these belief systems, the argument goes, outcompete those groups that fail to do so (Norenzayan 2013). Whether it is a genetic or cultural adaptation, the argument rests with the assumption that a belief in a moralizing, punishing god is the core reason for the success of these groups. In the next chapter, I will look at the evidence offered to support the hypothesis that human behaviour is strongly affected by a fear of god. For the remainder of this chapter, however, I will focus on the claim that the idea of god led to prosociality from a group selection perspective. To get a handle on group selection theory, I will first consider the notion of cultural evolution.

The debate

There seems to be no end to the debate over the application of cultural evolution, with some making a convincing case against even using the term (Dawkins 1982; Pinker 2012; Kundt 2015). Others say that evolution is not confined to genetic systems and can help explain virtually all the changes in human culture (Ridley 2010). Matt Ridley suggests that the flow of human culture is 'gradual, incremental, undirected, emergent and driven by natural selection among competing ideas' (Ridley 2010, p. 2). Ridley says that when we account for what happened in history, we place too much emphasis on design and planning and not enough on evolution.

It was Robert Boyd and Peter Richerson who argued that cumulative culture can be viewed with a Darwinian lens (Boyd and Richerson 1992). The different courses we take in fashioning our lives and the diverse institutions that we create are constantly changing, and the mechanism for the change, so the argument goes, is evolution. They suggest that the changes made are gradual, unintended and selective. These unintended cultural changes in turn have an impact on biological evolution in fundamental ways. For instance, in *Guns, Germs and Steel: The Fates of Human Societies*, Jared Diamond used evidence from biology, history, geography and anthropology to suggest that over the past 12,000 years, there has been a coevolution among humans, plant and animal domestication, and the structure of human societies. There is no question that genetic change has occurred due to cultural change. As pointed out quite clearly by Diamond, the ability to digest milk among people of European and east African ancestry was a genetic change responding to a change in the cultural environment between 5,000 and 8,000 years ago (Diamond 1997). Their ability to digest milk was due to the domestication of cattle for milk production. It was not that they were unable to invent dairy farming because they could not digest lactose; rather, the genetic change was a consequence of the particular culture. It is also fair to say that culture is cumulative, in the sense that humans have the ability to add new innovations to their already-existing knowledge without necessarily losing earlier customs. The main argument here, however, is that the BGP incorporate group selection to account for cultural evolution, and when doing so, it causes confusion because there is no consistency in their claims as to what is being affected by selective pressures in the environment. By looking at the differences between cultural and biological evolution, we will find that once the talk turns to certain religions outcompeting other religions, it becomes even more confounding. First I will consider biological evolution at the level of the gene.

Darwinian evolution

There are three essential features involved in biological evolution. First, the replicators (genes) must be able to make true copies of themselves. As Dawkins points out, 'natural selection favours high copying-fidelity' (Dawkins 1989, p. 17). Second, the competition is between genes competing with their alleles for survival, as the alleles are the rivals for their position on the chromosome of future generations. Finally, there will be random errors or mutations and, consequently, over a number of generations, there will be an accumulation of beneficial mutations within the population. As Radek Kundt points out, 'this process is mechanistic and the success of the replicators is determined solely by the number of their copies within the considered population' (Kundt 2015, p. 91).

Cultural evolution

One of the claims for using the term cultural evolution is due to the realization that much of the human world resulted from human action but not necessarily human design. For instance, Daniel Dennett and Matt Ridley are correct in claiming that much of consciously, planned design is responsible for far less of the design in the world than we tend to admit. However, Dennett chastises Steven Pinker because Pinker credits the comprehension of the intelligent designers for the improvements in cultural goods. Pinker writes, 'the striking features of cultural products, namely their ingenuity, beauty, and truth (analogous to organisms' complex adaptive design), come from the mental computations that "direct" – that is, invent – the "mutations", and that "acquire" – that is, understand – the "characteristics"' (Pinker 1997, p. 209). Surely, Pinker doesn't believe *everything* in culture can be accounted for by top-down design. However, is he not correct in acknowledging that many important cultural developments were the result of brave and bright intelligent designers? Is it not true that historically, some people have defended the rights of others by consciously 'directing' their movements with the 'understanding' that these people would have a better life? Perhaps Richard Webb's distinction between a 'special theory of evolution', which is Darwinian, and what he terms a 'general theory of evolution', which applies to such things as 'society, money, technology, language, law, culture, music, violence, history, education, politics, god and morality', would help us here (Ridley 2010, p. 5). The latter says that things do change, and they do so gradually but inexorably: 'they show trial and error; they show selective persistence' (Ridley 2010, p. 5). In this sense then, cultural evolution is a kind of natural selection.

Another way of thinking about cultural evolution is to say that cultural evolution proceeds by way of attraction, of which selection is only a part (Claidiere et al. 2014). It does seem odd to suggest that human culture consists of items that propagate through imitation when the historical evidence suggests that there was a great deal of sharing of information between cultures or groups. There are a number of different mechanisms involved in cultural propagation, and often they involve 're-production or recurrence rather than just reproduction' (Claidiere et al. 2014). Some information is no doubt lost with transmission, and it seems that cultural propagation is only 'partly preservative and partly re-constructive' (Claidiere et al. 2014). Having established some of the concerns that arise when using the term cultural evolution, I will now turn to the much more problematic claim that our propensity to cooperate with others can be explained as an adaptation to groups competing against each other.

Group selection theory

Proponents of group selection suggest that groups can be expressed as the adaptive unit, as there is a heightened competition between groups, and at the same time, there is less competition and fewer differences within the group. For David Sloan Wilson, in order to make the world a better place, there must be group-level functional organization (Wilson 2015). People must coordinate their activities in ways that achieve common goals. Wilson contends that human beings are evolution's latest major transition. He says, 'alone among primate species, we crossed the threshold from groups of organisms to groups as organisms' (Wilson 2015, p. 49). For Wilson, if we focus on functional organization, then religions can be seen as organisms that are collective units. This leads him to say that, 'religions *cause* people to behave for the good of the group and to avoid self-serving behaviours at the expense of other members of their group' (Wilson 2015, p. 79–80, *my italics*).

According to this theory, for a collective world view to be adaptive, it must be highly motivating psychologically and the actions motivated by the world view must outcompete the actions motivated by other world views (Wilson 2015, p. 87). Wilson says that actions inspired by some religions qualify as actions that proved good for the group and those religions outcompeted other religious beliefs and practices. He points to Seventh-Day Adventism and Mormonism as two religions that are very successful. Wilson writes, 'not only do Adventists famously keep themselves healthy, but there will be a jewel in their crown in heaven for every person whom they save' (Wilson 2015, p. 88). This gets confusing when we unpack the arguments for the success of religion from a cultural group perspective.

First, as was indicated earlier, trouble arises when trying to find consensus on the meaning of such ambiguous terms as religion. It becomes even more problematic when we talk about groups. For instance, Peter Turchin contends that 'the most important point is that the evolution of cooperation is driven by competition between groups. These groups can be teams, coalitions, even aggregations without any clear boundaries, or whole societies.' (Turchin 2016, p. 91). The line of thought outlined by Turchin is difficult to track because if there is no need to define the group, how do we establish what is outcompeting what?

To sharpen my point, consider a business owner who is accountable to his shareholders. Let's say this person turns to politics and is now running for president of the country, and during one of the presidential debates he divulges that, for a number of years, he has manipulated the system so that he has not paid one cent in federal taxes. One could say that by profiting himself and his shareholders, his dodging payment of taxes was good for the group. We can conclude from this scenario that groups composed of people who don't pay taxes outcompete groups of people who do pay taxes. In contrast, one can counter that when a billionaire doesn't pay his or her fair share of taxes, then this has a negative effect on the country, a much larger group. So from the same scenario, we can presume that not paying taxes was disadvantageous for the group, as those groups composed of people who do not pay taxes will get outcompeted by those groups who do pay taxes. If no boundaries are drawn, the idea of a group outcompeting other groups is confusing and, at times, misleading. From a Darwinian point of view, on the other hand, true replication must be precise.

In addition to the boundary problem in group selection, the argument rests with the claim that prior to the Holocene, the environment in which our hunting and gathering forebears evolved was rife with intergroup conflict (Bowles and Gintis 2011). Even if this was the case, how is group selection a better explanatory tool than gene selection or even individual selection? It is not as if the individual would have had a better chance of survival if he or she packed it up and went alone. All individuals engaged in social interactions are faced with very limited outside options, so they would be better off conceding to the terms of the interaction rather than leaving the safety of the group. In Chapter 3, I will argue that there were selective pressures to behave in a way to attract reliable partners as well as selective pressures to develop the cognitive capacities to find and assess reliable partners. The group is part of the environment in which conspecifics interact. It is not the group that evolves but the individuals within the group.

Third, in genetic evolution, genes must be capable of making true copies. In contrast, with group selection, the groups do not make true copies of themselves. Actually with the theory of group selection, there are no copies being made. The success of a gene is to make more copies, while the success of the group, such as the Seventh Day Adventists, is to expand the network by attracting or coercing more followers. As Steven Pinker points out, no one is making the claim that 'monotheistic religions are more fission-prone than polytheistic ones and that as a consequence there are more monotheistic belief systems' (Pinker 2012). The group then is not really a replicator but is essentially a super vehicle for vehicles that house replicators.

Fourth, in group selection, the mutations are not random. Changes are made in belief systems due to purposeful action. Peter Turchin says, 'there are many reasons why some ideas out-compete others. Some may appear to make more sense than an alternative, and become implemented after a *public discussion*. In other cases, prestige and fashion may play a role or it could be sheer chance.' (Turchin 2016, p. 226, *my italics*). This is fundamentally different from Darwinian evolution. There is no question that some ideas are better than others, but as will be argued later, the competition of ideas occurs within groups as well as between groups after reasonably assessing the alternatives. Rather than waiting to see what evolves culturally, often humans develop models to propose solutions to problems and then choose what appears to be the best option. Admittedly, at times we make stupid choices.

Fifth, the argument here is that groups are not as unitary or homogeneous as the group selectionists would have us believe. Tooby and colleagues argue that the group selection model is illusionary as it fails to take into account the evidence from the real world (Tooby et al. 2015). For instance, researchers generally model group cooperation 'assuming a social ecology of fixed groups, where agents interact in a single group for their entire life' (Tooby et al. 2015, p. 5). In the real world, the process of individuals cooperating within a group is often temporary and task-specific and 'forager band structure is frequently fluid, often involving temporary fissions, transfers and fusions' (Tooby et al. 2015, p. 6). Furthermore, ideas are not the intellectual property exclusive to particular groups; rather, they are shared among the people within the group as well as between groups. People within a particular group reflect on the cultural practices and at times are critical of the practices, and it is through dialogue and debate that some practices are changed. As Elliot Turiel points out, 'debate and argumentation commonly within *and* between groups are often at the root of moral and social transformations' (Turiel 2017).

Finally, Turchin contends, 'relative proportions of different beliefs are quite stable but they do change, given enough time. In other words, this cultural trait evolves. And that, remember is all evolution is.' (Turchin 2016, pp. 80–81). This is simply not true. As has been made quite clear by such advocates of evolution as Dawkins and Pinker, Darwinian evolution is more than simply change; it must be random, mechanical, algorithmic and unintentional. Some changes in society designed by humans, with the purpose of making the world a better place, are accomplished through reasoning and debating the issues. Sure there are unintentional consequences, but some of the biggest changes in society have been deliberate. And without some of these changes in society, humans would not be able to achieve 'thick cooperation'.

Take your mind back to post-war Britain and consider the brilliant mathematician Alan Turing. At that time, no one would have imagined that one day, individuals would be able to advise others about trivial events in their lives by simply updating their status on a handheld device and those people concerned about such superficial matters could make the pointless claim that they 'liked' it. Facebook is an unintentional consequence of the development of the computer. More importantly, consider that the use of social media by marginalized groups has led to an awareness of injustices by a greater number of people in the world, and this has led to a concerted struggle leading to greater freedom for more people. Turing didn't design the computer with either of these purposes in mind. This rationale was not something he was thinking about; rather, he was attempting to carry out mathematical and logical operations through algorithmic execution. The accumulation of knowledge and the use of that knowledge to design technology are, if anything, Lamarkian and not, strictly speaking, Darwinian.

Moreover, the scholars taking a group selectionist perspective inevitably conclude that we need to do a better job of designing our institutions. For instance, Joseph Henrich suggests 'once we understand humans as a cultural species, the toolbox for *designing* new organizations, policies, and institutions begins to look quite different' (Henrich 2016, p. 329, *my italics*). He goes on to say that 'humans are bad at *intentionally designing* effective institutions and organizations, though I'm hoping that as we get deeper insights into human nature and cultural evolution this can improve' (Henrich 2016, p. 331, *my italics*). Others suggest, 'seeing the Darwinian contest clearly can help us *create* social environments that favour altruism as the winning social strategy, improving the quality of life in anyone's hometown' (Wilson 2015, p. 116, *my italics*). Wilson also points out that 'future social arrangements need to be based more on intentional planning than ever before' (Wilson 2015, p. 114). He goes on to conclude that 'in our role as selection agents, we must function as altruists of the highest order' (Wilson 2015, p. 114). There is no question that we need winning strategies from intentional planning and we have to do a better job at designing our institutions, but we must also acknowledge that this is achieved through a slow process of argumentation. As argued earlier, changes happen in society, and often they are unintended but at times they are not; therefore, it is unclear how advantageous it is to use group selection to account for the development of a culture.

It appears that the group selectionists are confusing units of selection with development. Perhaps their distaste for a focus on the gene as the unit of selection leads them to a concern of genetic determinism. Yet the selfish gene theory doesn't lead to a disproportionate emphasis on genetic factors. When we talk about replication, we are concerned strictly with how genes replicate themselves. When the conversation turns to development, we must include nongenetic as well as genetic factors. When we say that we teach young children or that we 'design' our institutions or that we 'fight for the rights' of others, we are talking about development, not replication.

Even the reasoning behind the necessity of using group-level selection seems confusing at times. Group selectionists acknowledge that individuals display traits that are advantageous to both the individual and groups. They argue that sometimes the individual benefits at the expense of the group. The problem arises, so the reasoning suggests, when one tries to account for humans displaying traits that favour the group at the expense of the individual because of the risk of exploitation by free-riders, that is, those who benefit from the collective, yet have not themselves contributed. They argue that the only way to account for self-sacrifice and the fact that humans punish others at a cost to themselves is to see it through a group selection lens, since the threat of free-riders leaves unexplained how natural selection could have favoured its evolution.

To make his case, Peter Turchin, in *Ultra Society*, invites the reader to consider the dilemma faced by the main character Yossarian in the classic novel, *Catch-22*. Yossarian is struggling with the decision to fight a war, a situation where the odds are quite high that he may pay the supreme sacrifice, so he considers the more appealing option to freeride and leave the fighting to others. According to Turchin, Yossarian's reasoning is unassailable as 'a rational agent's best course of action is always to defect' (Turchin 2016, p. 58). There are a number of problems with this conclusion.

According to the group selection theory, free-riders such as Yossarian would have a fitness advantage over the other group cooperators on the platoon. If free-riders are left unchecked, then cooperation would dissolve. Turchin, and others who advocate group selection, assumes that the default design of individuals is to defect unless of course there are conditions put in place. Turchin defines rational agents as those who are motivated totally by greed and are utterly selfish. He says that 'it turns out that a group consisting entirely of rational agents is incapable of cooperation', and this has been proven mathematically (Turchin 2015, p. 57). Using Turchin's definition of rational agents as utterly selfish, then of course a group of them would not cooperate. However, the vast majority of human beings are not rational agents if we use Turchin's definition, or has there ever been, or ever will be groups of utterly selfish individuals.

The idea that rational individuals are selfish derives from the Rational Choice Theory (RCT), a popular view held by many economists and some political scientists. Essentially it is the belief that the primary motivation of human social behaviour is the pursuit of maximum self-interest. Proponents of the theory assume people make choices only with their own interests in mind and fail to take into consideration any other motivations of the person. Yet, many of the choices we make, on a daily basis, are made with others in mind, so it is difficult to draw an imaginary line between what can be defined as our own interest and the interests of others. A single-minded pursuit of one's self-interest when there may be a host of reasons for one's behaviour does not adequately capture what is meant by behaving rationally.

The economist and social justice theorist Amartya Sen argues persuasively that rationality of choice would be best explained by 'seeing rationality as conforming with reasons that one can sustain, even after scrutiny, and not just at first sight' (Sen 2009, p. 180). There is no question that dialogue is more consequential in public than in personal life, but viewing rationality in terms of presenting our decisions to scrutiny seems to be what being rational is all about. A person's choice is subject to a critical examination on reflection by the individual alone or in conversation with others. Sen points out that, in this way, we not only evaluate the decisions we make, given the values we hold or objectives we seek, but we can also critically investigate whether those values and objectives themselves are sustainable (Sen 2009).

In Chapter 3, I will explore the evolutionary history of hominins and suggest that the individual hominins who housed the genes that led to greater cooperation were more successful than those within the species who did not. 'Cooperation among small brained hominins in conflict with other scavengers became instinctual because it was pro adaptive to individuals; the genes that harboured behaviour that fostered these traits were evolutionary successful – they survived in the ancestral gene pool' (Coelho and McClure 2016, p. 70). Like our hominin ancestors, humans don't have an instinct to defect; rather, humans have a tendency to help and share with others as a result of the benefits that those traits provided in small-scale society. It will be argued that these 'thin cooperation' instincts evolved in small-scale groups that at times may be incompatible with 'thick cooperation', so our natural 'tribal mentality' sometimes needs to be held in check. One way to counter these biases is through freely arguing the issues.

Nevertheless, by failing to take into account the early evolution of our hominin ancestors, some scholars advocating group selection incorrectly assume that for a group of rational agents, the best course of action is to defect. It may be the case that it is rational to defect in the classical one-shot Prisoner's Dilemma (PD) games, but is it not also rational for two people to commit to cooperate with each other, if by working together they will have a greater chance of being successful? The suggestion that a rational agent will fake being sick to avoid combat until the person is tracked down by his or her enemies and killed doesn't make a great deal of sense if we take into consideration other motivating factors. I argue that a soldier who feigns sickness to avoid danger on the front line should be more worried about retribution from the surviving members within the group, rather than his or her enemies. I suggest that within a group of soldiers, any hint of not being willing to take one for the team would be met with subtle reminders of the group's disapproval of acts of cowardness. There is no group-versus-group competition going on, but there is indeed considerable social pressure within the group to conform and be seen as a good potential partner.

Rather than appealing to group selection theory to account for the actions of soldiers who willingly put themselves in danger for the benefit of the group, there is a much simpler rationale for their behaviour. I will show in Chapter 3 that humans are naturally motivated to cooperate; however, there are additional pressures and commitment mechanisms for why soldiers would cooperate rather than defect. It seems reasonable to treat intimate companions as honorary relatives. For instance, a group of soldiers commonly consider themselves brothers or sisters, a form of fictional kinship and take as self-evident that they will put themselves in danger to help their colleagues, knowing full well that their colleagues will not hesitate to do the same for them, a form of reciprocity. Of course, occasionally one finds free-riders, but I suggest that they are in the minority and historically they have been punished in some form or another by the actions of others within the group and not because of group competition.

There are two sets of theories that are posed to account for the free-rider problem. One group, and I include the BGP as well as Turchin within this category, suggests it is those groups who have established norms that maintain cooperation that outcompete other groups who don't have such norms. The alternative theory suggests that humans universally have the type of minds that leads them to easily cooperate in groups. The reason individuals have these predispositions is due to the fact that it was advantageous to the individual. The free-rider problem can be solved without group selection because more humans are wired to cooperate, rather than freeride.

Others such as Christopher Boehm argue that the reason people have strong and active consciences is due to the punitive social environment that prehistoric humans found themselves in. According to Boehm, there was social pressure to behave generously, and there was either discouragement or elimination of those who cheated or were bullies. He says that 'morality began with having a conscience and that conscience evolution began with systematic but initially nonmoralistic social control by groups' (Boehm 2012, p. 15). Boehm convincingly argues that not only the lives of individuals within small groups are influenced by the punitive actions of people within the group but the actions also shape gene pools in a similar direction (Boehm 2012). In our ancestral environment, food was scarce and risk to life was considerable, so it was beneficial for the individual to cooperate with others in the search for food and refrain from cheating.

While Boehm primarily points to the evolution of enforcement coalitions, Robert Frank suggests that humans evolved emotions that were motivationally powerful (Frank 1988). In Chapter 4, I will expand on the fact that our emotions are triggered by violations of trust, and this leads to ideas of fairness and injustice. We will also find that joint activity is emotionally powerful and that it is partly motivational because it is so rewarding (Tomasello 2014). In the example of the soldier contemplating defection, one must also take into consideration the environment that soldiers find themselves in. Consider that the shared activity and experience of the soldiers, often a life or death situation, make it an even more powerful trust-building mechanism. Turchin's suggestion that it is best for the individual to defect is far too simplistic as he fails to consider the internal and external commitment mechanisms that encourage the individual to cooperate.

Additionally, in the soldier example, what is the actual group that is successful? Is it the nation, the army or navy, the platoon, or a smaller group of soldiers within the particular squad? Evidence suggests that individual soldiers do not always care about the nation or larger groups but are particularly concerned about their small group of friends. As has been convincingly documented, the soldier is motivated to protect and to conform to the expectation of the 'primary group' (Shils 1951). Here again, the use of group selection doesn't help with assessing the evidence.

Furthermore, as has been argued, much change in society often originates within groups and not between groups. Let's consider Alan Turing again. Turing was a brilliant, gentle individual who assisted in breaking the German Enigma codes during the Second World War. To compensate him for his talent and his significant assistance in the war effort, the British Government offered him a choice between jail or, essentially, chemical castration. Now, the reason he faced jail had nothing to do with his loyalty to the country; rather, it was because he committed a homosexual act in private. Sadly, Turing chose a third option: he killed himself with a poisoned apple (Hodges 1983). Ironically, today one of the largest companies in the world employing many talented computer scientists is Apple. Nevertheless, the point here is that since 1967, homosexuality has no longer been illegal in Britain. How did this state of affairs come about? Did a group of homosexuals outcompete groups of heterosexuals? No, it didn't happen that way. Did groups that permitted homosexuality outcompete groups that did not? No, it didn't happen that way either. The reason why homosexuals are no longer jailed, castrated or stoned in many countries of the world is due to brave people fighting for the rights of other people and convincing others that persecuting individuals because of their sexual preference and to harass people who engage in consensual sexual acts is based on ignorance and hatred. One of the main arguments in this book is that we have greater cooperation in the world today because, historically, brave people within groups have fought for the disadvantaged and were able to develop intelligent solutions to complex matters and were able to convince others of their cause. These examples of people fighting on behalf of the disadvantaged are relatively recent. To accept a person who traditionally is seen as an outsider requires the recognition of that person's individual autonomy, a prerequisite for 'thick cooperation'. I will have more to say about individual autonomy in the following chapters.

Even if we look at the beginning of state formation, the idea of groups bonding together because of some type of unifying cultural adhesive doesn't fit with the historical evidence. First, it is very difficult to give a precise date as to when the political landscape becomes dominated by a state. Consider that until about 400 years ago, 'one-third of the globe was still occupied by hunter-gatherers, shifting cultivators, pastoralists, and independent horticulturalists' (Scott 2017, p. 14). James Scott reminds us that despite the standard mythology of a central state, it is important to keep in mind that for many years after its first appearance, the state was not constant and was in fact quite fragmented.

What does seem necessary for establishing the first states was coercion. For instance, historians have suggested that the Great Wall of China was necessary to keep the people in as it was to keep the barbarians out (Lattimore 1962). James Scott says the evidence for the 'extensive use of unfree labour – war captives, indentured servitude, temple slavery, slave markets, forced resettlement in about colonies, convict labour and communal slavery is overwhelming' (Scott 2017, p. 29). With the Sumerians, Babylonians, Egyptians and Assyrians, we find a world 'in a very profound sense, a world without free men, in the sense in which the West has come to understand the concept' (Finley 1959, p. 164). The evidence provided by Scott shows clearly that it would be difficult to deny that bondage, forced labour and slavery were the means by which states were able to expand the grain-labour module, which was at the foundation of the natural state. If Adam Hochschild's claim that as late as in 1800 approximately 75 per cent of the world's population could be considered as living in bondage is correct, then the group selection model offered by the BGP doesn't seem to make a great deal of sense (Hochschild 2005). Rather than suggesting that groups outcompete other groups, I suggest that the change from 'thin' to 'thick cooperation' required an environment that encouraged freedom and a recognition of the autonomy of the individual. I will explore this claim in greater detail in Chapter 4, but for now I will turn to the problems that arise from using group selection to make the dubious claim that groups consisting of god-fearing people outcompeted other non-god-fearing groups.

As has been argued, the cultural group selection framework has been used to argue that belief in gods or supernatural agents is adaptive as it united people in such a fashion that it allowed religious groups to outcompete other groups. By looking at how this arguments unfolds, we will find that it causes a great deal of confusion. For instance, in *Supernatural Selection*, Matt Rossano asks whether we are better off with or without the idea of god or supernatural agents (Rossano 2010). Rossano reasonably suggests that shamanism is practiced in one form or another in nearly every traditional society. Just as I have argued in this chapter, Rossano suggests that one of the core features of shamanism is the fact that an ASC is used by some members of the society to connect to a spiritual world. He notes that once humans evolved this capacity, there was a potential for shamanism (Rossano 2010, p. 72). He also notes that shamans were selected from the most eccentric individuals in the community. So far, so straightforward.

According to Rossano, the shamans were able to identify the discreditable members of the community who had the unacceptable habit of making the gods angry. By having such talent, the shamans were seen as authority figures who ensured group norms and decisive individualism (Rossano 2010, p. 73). According to this view, prior to shamanism, individuals were free to sin, but with the prescience of the shamans, misbehaviour was brought to light. Furthermore, by coupling ceremonies with the supernatural element, rituals became even more effective at securing group solidarity. The groups who were psychologically united by powerful rituals would tend to share resources, providing 'insurance against starvation, giving them a fitness advantage over *purely secular groups*' (Rossano 2010, p. 93, *my italics*).

Rossano goes on to suggest that the introduction of the supernatural acted as a binding mechanism that harnessed together a number of cognitive capacities, namely, the capacity for engaging in ritual, capacity for ASC, ability to form beliefs and the ability to form social bonds, directing them towards the single adaptive purpose of healing. Rossano calls this 'Placebo consolidation' (Rossano 2010, p. 158). Supposedly, the emergence of shamanistic healing,

70,000 years ago, gave these religious communities an adaptive advantage. Not only did religion offer health and healing advantages but it also promoted a deeper commitment to group norms. For Rossano, this made 'religious groups more competitive than *non-religious groups* when there was direct conflict over resources' (Rossano 2010, p. 167, *my italics*). By adding the spiritual element, religion 'directs gut level moral sentiments in specific ways, encouraging collective interest over self-interest and thus facilitating effective group action' (Rossano 2010, p. 176). For Rossano, this is true today just as it was in our Palaeolithic past (Rossano 2010, p. 171).

There are a number of problems with this argument. First, as Lewis-Williams makes clear, 'shamans cannot heal broken legs' (Lewis-Williams 2010, p. 125). Lewis-Williams also points out that the possible merits of a placebo effect for religion are not being doubted here; however, one must remember that the healing supposedly administered by the shamans is largely due to the stories that they had created in the first place. It is not entirely clear whether the stories made up by the shamans were more beneficial to the group or the individual shamans. Surely it must be acknowledged that some of the stories told about the spirit world, not just by shamans but by leaders of all religions, can potentially cause more distress in the lives of people than provide comfort. Things can go terribly wrong if you happen to be one of the targets of the shaman's story. Consider some of the unscrupulous stories told about homosexuals by many religious leaders.

Second, the group selection argument rests on the notion that groups with favourable traits outcompeted groups with less-favourable traits. With Rossano, as well as others who argue from a group selection perspective, groups of people made up of god believers outcompeted groups made up of non-believers. The problem here is the lack of evidence for any groups consisting solely of non-believers. I have argued, and the historical evidence seems to support my claim, that for most of human history, everyone believed in the gods or supernatural agents. In these shamanistic communities, it is not as if they considered themselves practising a particular religion. There were no communities where the people gathered for rituals on Sundays while others contemplated the pros and cons of the string theory. Rather, people took it for granted that their myths described the way of the world. Even today, in our secular societies, there is no evidence for cultural groups made up solely of non-believers. To make the group selection argument 'work', proponents resort to fictitious groups of non-believers. If we are going to use this logic, one could argue that the reason that the Catholic Church consists solely of male priests is due to the fact that those groups consisting exclusively of male priests outcompeted all the alternatives.

Justin Barrett also plays the group selection card. He argues that monotheism that can be 'traced back to Abraham enjoyed very special selective advantages over most other religions' (Barrett 2004, p. 88). He goes on to say that 'many local religions folded (or are folding) under competition from other faiths and from scientific and philosophical inquiry in part because their ritual systems lack the conceptual control to survive changing demand of validity' (Barrett 2004, p. 88). Finally, he proposes that 'shamans, traditional healers and other spiritual leaders have had to learn about modern medicine to retain their positions of religious leadership' (Barrett 2004, p. 88). Although Barrett doesn't elaborate, apparently the shamans needed career development training to upgrade their knowledge and skills to keep up with the emerging technology. There does not appear to be any outcompeting of groups going on here; rather, it appears to be an attempt to use evolutionary theory to argue for the merits of monotheistic religions. The argument of certain religions outcompeting other religions and/or fictitious atheist communities is not comparable to Darwinian random mutation and selection by any stretch of the imagination. This argument is confusing and unnecessary for an explanation for the persistence of a belief in gods and the supposed selective advantage of such belief.

Skyhooks or cranes?

Notwithstanding his sweet tooth for memes, Daniel Dennett consistently hits them out of the park with his intuition pumps, ratherings, *free-floating rationales*, zombie hunches and the odd deepity. In his finest book *Darwin's Dangerous Idea*, Dennett suggests that there are only two ways to account for the many ways that an organism can be said to be alive. It comes down to either a miraculous Intelligent designer or all the design work lumbered along by the Darwinian notion of descent with modification. Dennett cleverly uses the metaphor of skyhooks and cranes to argue that the process of evolution is like a crane, a mindless, motiveless, algorithmic process. On the other hand, he defines skyhooks as 'miraculous lifters, unsupported and unsupportable' (Dennett 1995, p. 75). Dennett wittily argues that the slow process of evolution is due entirely to cranes that do all the heavy lifting in Design Space to create every organism on earth by taking baby steps and eventually, with enough time, cross huge distances.(Dennett 1995, pp. 75–76). By using the explanation of evolution by natural selection, Dennett argues that there is no need for the miraculous skyhook. The questionable evidence offered by the cultural group selection authors in support of their claim that religion was an adaptive advantage for our species is, it seems to me, an attempt to use the logic and language of cranes to justify the fitness advantages of a particular skyhook.

Summary

I began this chapter by differentiating two systems involved in human cognition. One process is spontaneous and reflexive and serves the goals of the genes, while the other system includes declarative and reflective thinking that serves the goals of the individual. The evidence from CSR researchers suggests that the idea of god is derived from the reflexive assumption about domain concepts such as a person or artefact, coupled with counterintuitive properties. Although these ideas are universally held, they are not necessarily inevitable. I suggested that for the idea of gods to spread, stories had to be told about them, and the evidence suggests that it was shaman-type individuals who had superior proficiency to tell such stories. I went on to suggest that some scholars argue that since TASS has served us well in our evolutionary history, we should cherish our intuitions. Some even made the implausible suggestion that this is where we should get our values from (Barrett 2004; Hood 2012). I argued that we do not get our values in this way, nor should we, and that while we have unintended consequences, we must reason about complex issues in order to expand human cooperation. Humans are not the emotional dogs with rational tails suggested by scholars such as Haidt, but rather, keeping with the metaphoric language, we are rational dogs, and at times, we are influenced by our emotional tales. Finally, I argued that the evidence supporting the claim that groups made up of individuals who believe in god outcompeted groups made up of non-believers is fraught with difficulties. In the following chapter, I look at the evidence that suggests that it was the idea of a punishing god that reduced selfish behaviour thus leading to large-scale cooperation. We will find that the evidence benchmark to support this claim has been lowered to such a depressed level that it is not worth taking seriously.

2

A modest proposal

In the previous chapter, I argued that by virtue of how the human mind operates, we have the capacity to believe in supernatural agents or gods and, consequently, be influenced by the stories told about them. In this chapter, I will assess the evidence that purportedly supports the claim that the idea of God provided a fitness advantage to humans. Some have suggested this belief provided an advantage at both the individual and group levels, and subsequently, humans evolved to fear supernatural agents (Johnson 2016). There are two related explanations as to how monitoring by supernatural agents leads to an increase in prosocial behaviour and thus provides a fitness advantage. Punishment Avoidance suggests a fear of God evolved at the individual level as it prevents individuals from defecting and incurring the costs of punishment in this life or the next (Johnson and Kruger 2004). It has been suggested that this is primarily about how such a fear 'emerged' in human evolution, sometime in the Pleistocene, when one's reputation became threatened by the human capacity for ToM and the acquisition of complex language skills (Johnson 2016, p. 204). Alternatively, Cooperation Enhancement is explained at the level of the group, as groups of individuals composed of individuals who cooperate outcompete groups made up of fewer cooperative individuals (Wilson 2002). Ostensibly, this took place within the last 10,000 years when a fear of supernatural punishment 'developed' in larger societies when the reputation of strangers could not be so easily tracked (Johnson 2016, p. 204). The two mechanisms can be combined to suggest that religions with followers whose selfish behaviour is curbed by a belief in moralizing, punishing gods will outcompete other groups composed of individuals who do not believe in such gods (Norenzayan 2013). In the previous chapter, I outlined the reasons why the group selection theory is problematic, particularly when it comes to suggesting that religious groups outcompeted nonreligious groups in human evolution. Nevertheless, the BGP suggest that it is this belief in moralizing, punishing Gods that caused human societies to scale up from hunting and gathering groups to the large complex civilizations we find today.

Regardless of whether religion was a product of genetic or cultural evolution or even a combination of both, the BGP claim that the enabling element that 'pushed human groups to rapidly scale up from hunter gather origins to the vast societies of millions today' was the belief in punishing, moralizing supernatural agents (Norenzayan 2013, p. 75). In order to support their thesis, the BGP reason that if a belief in supernatural agents 'causes prosocial tendencies then experimentally induced thoughts of God should increase prosocial behaviour in anonymous situations' (Norenzayan 2013, p. 34). Indeed the argument rests, for the most part, on a number of game theory experiments.

In *The Origin of Species*, Charles Darwin boldly asserted that if it could be demonstrated that any complex organ could not have evolved through numerous, slight modifications, then his theory would break down (Darwin 1859, p. 189). While no one has provided evidence to seriously challenge Darwin's profound insight, in this chapter, I will demonstrate that the evidence offered by the BGP has too many big holes to support their big claims. The purpose of this chapter then is to examine critically the evidence provided by the BGP and assess whether there is any reason to believe that a person's belief in a punishing, moralizing God leads to prosocial behaviour. Without this evidence, I suggest that the BGP hypothesis breaks down.

A Modest Proposal

In 1729, the celebrated satirist Jonathan Swift published the lengthy essay titled, 'A Modest Proposal for Preventing the Children of Poor People from Being a burthen to Their Parents, or the Country, and for Making Them Beneficial for the Public'. Swift's acrimonious essay was in response to the deplorable living conditions of the people living in Ireland. Swift sarcastically offered a way to convert the starving children in Ireland into something useful for society. His modest proposal was to beef up the children in order to feed them to the rich landlords. According to Swift, this would benefit the overall economy of Ireland and would solve the social and economic problems of the poor better than any other means that had been suggested.

Although the hypothesis that human behaviour is strongly influenced by the imminent threat of God, who will reward or punish human conduct either in this life or the next, has been argued by pointing to a number of scientific experiments, the main course of reasoning comes served with Swiftian sides. Consider the suggestion that trial by ordeals 'might have *cleverly* determined who was guilty and who was innocent with reasonable accuracy, without need for actual divine intervention' (Norenzayan 2013, p. 13 *my italics)*. Norenzayan goes on to suggest that what made the ordeals function *effectively* was a genuine belief in divine intervention. He says the ordeals may appear strange, but the critical feature 'that made it work' affects billions of people around the world (Norenzayan 2013, p. 14 *my italics*). This critical feature the author is speaking about is divine social surveillance, and a belief in such supernatural agents or gods, so the argument goes, caused a 'sea change' in human cooperation (Norenzayan 2013). To be fair, Norenzayan does admit that we can't, for certain, be sure that the ordeals were effective, but my question is, how can one even consider that trial by ordeal was effective or a good procedure for determining guilt? A reader could be forgiven for thinking that this line of reasoning is satire, as it is not entirely clear whether the author is being serious when suggesting that a threat of such extreme punishment is what keeps people in line.

Rather than suggesting trial by ordeal may have been effective, a much more reasonable question to ask is, why did it take so long to discard a procedure that seems so insanely unfair and foolish? In putting the matter in this way, I may be criticized for a sense of superiority over my medieval ancestors; however, to investigate how we have come to a point where we regard trial by ordeal as wrong or ludicrous is a legitimate and reasonable question to ask. I will pursue this question in the following chapters, where I will argue that much of the progress that led to 'thick cooperation' among large groups of people stemmed from a bottom-up process, where small groups of individuals campaigned for the welfare of others. It was the action of these brave people in history that gave a greater voice for the disenfranchised people in society.

It is not just trial by ordeals that has Swiftian overtones, but there are a number of other nonsensical conjectures made in support of the Supernatural Punishment Hypothesis (SPH). In questioning whether police or governments are as effective as God in deterring selfish behaviour and encouraging cooperative behaviour, one author notes that in most Western countries people are dissuaded from crime *only* by fines or time in prison (Johnson 2016, p. 228). Apparently this is unacceptable, as these punishments are *pathetic* compared to eternity in hell (Johnson 2016, p. 229, *my italics*). The author claims that 'God's unrivalled powers of punishment may remain a much more powerful deterrent than anything secular institutions can come up with' (Johnson 2016, p. 228). Can this be a serious claim in contemporary society? I argue that the form of punishment imposed on individuals convicted of crime in the majority of Western countries is, for the most part, not pathetic, but civilized. Are humans such a pathetic lot that we need to be scared straight to cooperate? Is it not enough to face moderate punishment rather than be threatened by such extreme intimidation? Later in this chapter, I will explore the evidence offered by criminologists refuting the claim that the threat of cruel punishment reduces crime.

Even the general claim that supernatural monitoring or supernatural punishment strongly influences cooperative behaviour seems illogical. Consider the following anecdote, found in *God Is Watching You*, that Johnson says makes clear why 'supernatural reward and punishment remains essential' for human cooperation (Johnson 2016, p. 10). Purportedly, in the 1960s, government officials in the City of Amsterdam came up with the idea of leaving bicycles around the city for use by local residents. Unfortunately, even in a relatively low-crime city like Amsterdam, the bikes were eventually stolen. Fast forward to 1993, where a similar campaign was launched in Cambridge and we find the same circumstances, but the results were even worse, as 300 bicycles were stolen on the very first day. Today, many urban centres have comparable programmes, but they come with some type of payment and registration system in order to keep track of the bikes in the event one happens to go missing. The author noted that it took a long time for the authorities to realize that what was needed was an effective system of payment and a threat of punishment in order to effectively dissuade citizens from stealing the bikes (Johnson 2016, p. 13). That is a sensible observation; however, the registration system was initiated in order that any person who does not return a bike will be apprehended and fined by other people. There was no mention of supernatural agents. In fact, if people were so concerned about punishment by supernatural agents, then there would have been no need for bicycle registration in the first place.

The central claim of the SPH is 'supernatural punishment is an extremely efficient deterrent because it allows constant and pervasive monitoring, severe and unlimited punishment, and shifts the burden of administering sanctions from people to supernatural agents' (Johnson 2016, pp. 95–96). While this may appear to be a cost-effective, infallible and highly efficient mechanism, the hypothesis rests on very thin ice. A common aphorism used by lawyers and judges in criminal trials and first articulated by Lord Hewart in 1923 is not only must Justice be done but 'should manifestly and undoubtedly be seen to be done' (Sen 2009, p. 393). This adage applies fittingly to both the Supernatural Monitoring Hypothesis (SMH) and the SPH, as once a person cheats or steals or doesn't hold up to his or her end of a bargain, without suffering any observable repercussions, the person, as well as others, will soon discover justice is not going to be done. Realizing that the gods are not everywhere or infallible will lead to the same cheater problems that the belief is suppose to prevent, unless real-life mechanisms are put in place to ensure the guilty are punished. Examples of this type of problem occur time after time. For instance, school administrators have anti-bullying campaigns and special events discouraging such behaviour; however, if a child complains about being harassed and the administrators fail to ensure consequences for the offender or take seriously the complaints made by the victim, then the inappropriate behaviour of the bully will continue.

Burden of proof

Since the purpose of this chapter is to critically assess the evidence provided by the BGP, I should make clear what I mean by the term evidence. In a criminal investigation, the detective or group of detectives gathers evidence in order to determine whether a particular event or state of affairs was caused by the conduct of the accused. In order for an accused person to be convicted of a criminal offence in Canada, the prosecution must prove the allegation beyond a reasonable doubt; however, the police, to lay a charge, need only reasonable grounds. The more evidence the police gather, pointing to the guilt of the accused, the more likely the accused will be found guilty of the charge. It is not being suggested here that the burden of proof required to support the BGP claims requires the same level as found in criminal court. Certainly, the presumption of innocence is at the heart of the judicial system, so a very high standard of proof should be placed squarely on the shoulders of the Crown. Without a doubt, it is a big claim to suggest that the belief in a punishing god facilitated the rise of human civilization even though no one's liberty is in jeopardy. I am not even suggesting that we use the less potent 'on a balance of probabilities' test required in civil court. Rather, I am suggesting that the evidence should be assessed in order to see if it is sufficient to lead a reasonable person to be persuaded by the BGP claims, and I will argue that the evidence, to date, provides no air of reality to their assertions.

Let me give an example to demonstrate where I believe the level of evidence offered by the BGP rests. Consider a sexual assault investigation, where a woman reports to the police that she was sexually assaulted by a male suspect. Imagine that the only evidence found at the scene of the crime was a ripped piece of green material that most likely came from the suspect's shirt during the assault. This has very little evidentiary value, as many men have green shirts. Now consider the forensic investigators discover the fibres from the ripped piece of green material are consistent with the type of shirts worn by members of a local boys private school. The evidential value of that shirt has increased, as the number of suspects has been reduced. Suppose further that a hair was located on the ripped piece of material and the forensic investigators now are able to determine the colour of the hair. Once again this would reduce the pool of suspects. Finally, suppose the hair seized came intact with the follicular tag, the part of human hair where the richest source of DNA is found. This would be optimal evidence, as it provides a causal connection among the suspect, victim and crime scene. I will argue that the evidence offered in support of the BGP hypothesis is equivalent to the ripped piece of green material – essentially useless. I will now turn to the experimental evidence that is often cited in support of the Big God hypothesis.

Experimental games

Much of the evidence to support the claim by the BGP comes from experimental studies using the priming technique. Behavioural priming is a procedure used by social psychologists where the researchers expose stimuli, such as words or pictures, or a set of stimuli, such as sentences, to one group of participants who are required to unscramble the words, while a second group of participants are not exposed to such stimuli. The prime purportedly activates a concept which influences the behavioural response of the primed participants. Ostensibly the person is unaware of the connection among the prime, concept and behavioural response, and this lack of awareness is the reason researchers believe that the response is automatic and without reflective assessment. Indeed researchers go to great lengths to try to keep their subjects in the dark.

In one classic priming experiment, researchers asked participants to select the word that didn't belong from a number of scrambled words, and when rearranged, the words formed a coherent sentence (Bargh 1999). The word that did not fit into a sentence was one that was related to the concept of being old. The participants were apparently unaware of this relation. The researchers found that those participants who were exposed to the old-age words left the laboratory at a slower walking speed than the participants who were not exposed to such words. The researchers argued that this demonstrates that human behaviour can be influenced by priming without the person's conscious awareness.

Increasingly, social psychologists have used this priming tool to investigate human behaviour. For instance, in one study, participants were shown pictures that were related to business concepts, such as a briefcase. The researchers, using word completion tasks, found that the participants who were exposed to such business-related primes were more likely to use words related to competition rather than cooperation (Kay 2004). In another study, participants who were wearing heavy backpacks were more likely to judge a hill as steeper (Bhallam and Profit 1999). Other studies suggest that participants who were primed with cheerleader stereotypes performed poorly on problem-solving tests (Galinsky 2008), and yet others suggest that females who were simply reminded about their Asian American ethnic background performed better on a math exam than those who were asked questions about their gender (Shih et al. 1999). It is clear that the priming literature is quite diverse, yet it is not quite as clear whether it is an effective tool in predicting behaviour.

Priming critics

Recently, there has been criticism of the priming methodology (Doyen et al. 2012; Shanks et al. 2013; Newell and Shanks 2014). Arguably, the main problem is a failure to replicate such studies. For instance, in a study by one group of researchers, participants were tasked with describing the characteristics of a typical university professor, while a second group was asked to describe a soccer hooligan (Dijksterhuis and Van Knippenberg 1998). Both groups were then given a general knowledge task, and the first group of researchers found that the participants primed with the professor characteristics performed better, while the hooligan-primed group performed worse. In a separate study, a second group of researchers attempted nine similar experiments with 475 participants in order to replicate the original study; however, they were unable to obtain a positive intelligence priming effect (Shanks et al. 2013). Similarly, in another attempt by researchers to replicate an original study, Stephanie Doyen and colleagues failed to replicate the Bargh experiment (Doyen et al. 2012). The point here is that we have to be cautious in making claims from individual studies, especially when few participants are used. We will find that some of the 'ground breaking' experiments from the BGP include very few participants, and yet from a number of artificial and contrived tasks, the researchers propose embellished claims about human psychology and behaviour.

A second criticism of the priming methodology is the difficulty involved in ascertaining whether the participants were aware of the primes. For instance, in one study, it was found that the people who were primed with scrambled sentences were fully aware of the social category with which they were being primed (Doyen et al. 2012). The argument here is that the participant may be able to figure out what is going on and hence be tempted to conform to the experimenter's expectations. Furthermore, Doyen points out that there have been a number of studies suggesting that the expectation of the experimenter can influence the behaviour of the participant (Dehaene 2006; Doyen et al. 2012).

A third criticism, and this pertains particularly to religious priming, is the difficulty in ascertaining what type of effect a specific prime has on behaviour. There could be a host of potential effects that can emerge from an activated concept. That is, constructs, when activated, can increase the accessibility of other constructs linked in one's memory. For instance, a song from a favourite musician may prompt a memory of a particularly good time with a significant other when that tune was first heard. Furthermore, memory is fragile, so we have to be cautious when drawing conclusions from word-priming experiments. To demonstrate the fragility of human memory, psychologists have used a number of tests. For instance, look at the words in Figure 2 for ten to twenty seconds and try to memorize as many words as you can.

bed, rest, awake, tired, dream, wake, snooze, blanket, doze, slumber, snore, nap, peace, yawn and drowsy.

Figure 2 Memory Test.

After looking at the words, close the book and take a piece of paper and write down as many of the words as possible that you believe were on the list. If you are like most people, you would have jotted down between six and ten words, and there would have been a good possibility that you wrote down the word sleep. In fact, researchers have found that about 40 per cent of the people believe that the word sleep was one of the words on the list, even though it wasn't (Kahneman 2011). We will see that in the religious priming literature, participants are primed with ambiguous religious terms, such as 'spirit', that could mean a number of different things to different people, yet the researchers conclude that this priming demonstrates that the participant was made aware of a morally, punishing supernatural agent (Shariff and Norenzayan 2007).

Despite the cautionary flags raised in placing too much emphasis on the priming literature, some researchers advocate that 'religious priming provides a powerful tool to test the *causal* effects of religious thinking in theoretically relevant psychological outcomes' (Norenzayan et al. 2016, *my italics*). In a recent meta-analysis, the authors suggest that their investigation provides support for the effect of religious priming on prosocial behaviour for religious participants.

The 'religious participants' qualifier is important because in all the studies examined, there was 'no reliable effect whatsoever for non-religious or low religious participants' (Norenzayan et al. 2016). That is, there is no evidence to suggest that there is a difference between believers and non-believers in prosocial behaviour. Nevertheless, the BGP still maintain that it is a fear of God that stimulates such prosocial behaviour. I argue that by examining the religious priming literature, it will be made clear that there is a much more plausible explanation for prosocial behaviour other than the influence of a belief in Big Gods, and it will be found that the idea of God has nothing to do with it.

The meaning of prosocial behaviour in broad terms relates to economic measures that are related to ethical, cooperative, or generous behaviour or attitudes. These behaviours include cooperating in the PD game, generosity to strangers in the Dictator Game (DG), generosity to the common good in Public Good Games (PGG), refraining from cheating or lying and willingness to volunteer services (Norenzayan et al. 2016). Most of the studies to date involved priming the participants with 'religious' terms. Yet recently, there have been some studies focusing on the contextual setting instead with the intent to evoke 'religious' concepts.

Supernatural punishers

The BGP have made some daring claims about religion considering the evidence provided in support of their claims is less than persuasive. For instance, Dominic Johnson maintains that human behaviour is 'strongly influenced by the idea that God or a supernatural agent will reward or punish our actions' (Johnson 2016, p. 7). He cautions that without a belief in supernatural punishers, we may be 'playing with fire' (Johnson 2016, p. 233). We will find, however, that there is very little evidence, if any, to suggest that it is the idea of God that influences prosocial behaviour. These claims about the influence of belief in God on human behaviour have been modified slightly because some researchers now concede that those claims do not mean that religious priming alone encourages prosocial behaviour or that religion inspires prosociality indiscriminately or universally. The BGP also acknowledge that the religious prosociality is 'most likely' directed towards in-groups and that religion is 'not the only or most effective or most desirable way toward achieving greater cooperation' (Norenzayan et al. 2016). Nevertheless, there still is the BGP claim that without the idea of punishing Big Gods, 'ancient Egypt – as well as nearly every other large-scale society in history –

wouldn't have been able to get off the ground' (Wade 2015). Author Lizzie Wade quotes Edward Slingerland as saying all-knowing Big Gods are 'crazily effective' at enforcing norms since 'not only can they see you everywhere you are, but they can actually look inside your mind' (Wade 2015). There is no question that aspects of religion encourage cooperative behaviour within the group, and there is no question that many religious people of all faiths are exemplary individuals, but one of the questions to be answered in this book is whether it was a fear of god that made them so. The claim that it was specifically a moralizing, punishing god that led to greater cooperation and that this consequently led to large-scale civilization is greatly exaggerated. I will now turn to an examination of the priming evidence.

Religious priming evidence

I am not suggesting here that all the priming literature is useless; however, as the aforementioned concerns suggest, it is very clear that there are limitations to such experimentation and therefore caution is warranted. Joseph Cesaro argues that the main reason why the priming literature is currently under threat is due to it being a relatively new method (Cesaro 2014). He says that researchers who resort to this type of technique should show more methodological rigor. It doesn't appear, however, that all researchers heed this advice. For instance, some of the BGP claim that experimenters can take subjects to the laboratory to 'test whether experimentally manipulating beliefs in supernatural punishment affected their behaviour under controlled conditions, so that *causality* can be examined and confounding factors can be ruled out' (Johnson 2016, p. 64, *my italics*). The claim here is quite clear – through controlled experiments, researchers can decipher whether a belief in a punishing god causes individuals to curb their selfish behaviour and consequently promote prosocial behaviour. By examining these 'groundbreaking studies', we will find that the evidence presented by the BGP is far too weak to support their bold claims.

Azim Shariff and Ara Norenzayan hypothesized that people primed with supernatural beliefs would be more generous than those who were not primed. To determine generosity, the experimenters used the DG. This game was first introduced in the mid-1980s and is often used by economists to understand the decisions people make (Kahneman 2011). In the DG, player one is given a certain amount of money by the experimenters and is asked to give any portion of the money to player two. It is called the DG because the decision lies solely with player one; player two has no say in the matter. If individuals were only concerned with maximizing their own profits, they would keep all the money and give the other player nothing. After all, there are no consequences for player one, as this is a one-time affair and entirely anonymous. Typically these studies have shown that people tend to give away 20 per cent of the money. However, it is not uncommon for people to generously give up half the money (Almenberg 2013). Nevertheless, let's look at the experiments that allegedly offer causal explanations for the link between a punishing god and prosociality.

In one study, 'God Is Watching You', participants were randomly assigned either to a group of individuals primed with religious terms or to a group who was not primed with religious terms. The religious-primed group was given 10 five-word sentences which included such words as god, spirit, prophet, sacred and divine. The participants were asked to unscramble the words to make a sentence. For instance, the words 'felt she eradicate spirit the' would become 'she felt the spirit' and the words 'dessert divine was fork the' would be unscrambled to read, 'the dessert was divine' (Shariff and Norenzayan 2007, p. 804). After the priming session, each participant played a one-shot version of the DG with a 'confederate'. 'Confederate' is the term used to describe a player who was actually a member of the research team. The researchers found that the participants who were primed with religious words gave more than twice as much money as the people in the control group. For Johnson, this 'groundbreaking study, demonstrates that something as intangible as supernatural beliefs may be a serious promoter of cooperative behaviour' (Johnson 2016, p. 65).

One major problem with this experiment is that the target words used for religious priming are quite ambiguous and are open to multiple interpretations. The authors of this study, and other BGP, who draw on it as evidence, use the terms religious, supernatural or god concepts interchangeably, yet words associated with religion may be packaged into different groups (Ritter and Preston 2013). For instance, words such as God and Prophet are concepts that may be related to supernatural agents, while spirit or divine makes up religious adjectives, and prayer and worship are related to religious practices. Other words, such as Church, Mosque or Temple, have institutional aspects. In fact, some studies suggest that priming subjects with words such as heaven and miracle increased the intention to help others, while words like Bible and chapel did not (Pichon et al. 2007). This suggests that it is problematic to conclude what these words actually mean to different people.

Furthermore, two of the five words used in the Shariff and Norenzayan study are not necessarily 'religious' words. The word spirit, for example, may be associated with cooperation, as in 'team spirit', or encouragement as in 'that's the spirit'; depression as in 'out of spirit'; alcohol beverages as in 'wine and spirits'; or simply one can associate spirit with an angel or fairy. These words bring to mind a vast number of psychological associations, so it is unclear what the underlying mechanism is that produces the behavioural effect described. It is a stretch of the imagination to suggest that 'she felt the spirit' or the 'dessert was divine' summons punishing gods to the minds of participants. Making decisions, determining how to behave and contemplating what to contribute in economic games in an institutional setting come from a myriad of conscious and unconscious processes. The argument is not that these religious primes may activate generic concepts about religion but to suggest that general religious terms elicit big God concepts, particularly punishing, a moralistic omniscient God, seems wishful thinking.

From this study, the authors jump to the conclusion that the 'activation of God concepts, even outside of reflective awareness, matches the input conditions of an agency detector and as a result, triggers a hyperactive tendency to infer the presence of an intentional watcher' (Shariff and Norenzayan 2007, pp. 807–808). According to these authors then, tasty desserts and feeling the spirit prompt one to feel as if they are being watched by supernatural agents, which in turn curbs selfish behaviour. This seems to be one giant leap of faith as any number of words, including secular words, can be associated with self-awareness. The terms need not be religious, let alone associated with punishing gods. From these two studies, the authors wildly speculate that early religion likely facilitated population growth, and the evidence from this study suggests that the 'invocation of supernatural agents may have played a central role' (Shariff and Norenzayan 2007, p. 808).

To demonstrate the problem with identifying particular triggering mechanisms in the priming research, one author suggests the following example (Hogarth 2014). Imagine that you are a social psychologist interested in seeing whether there is a connection between romantic attachment and lighting in restaurants. You could hold an experiment by randomly assigning volunteers for a blind date to have dinner either in a well-lit room or in one with the lights dimmed. You reason that the room with the darker atmosphere would encourage romantic attachment. There is only one variable, and it is environmental, but by manipulating it, you may be able to determine whether it had any significant effect.

In this hypothetical romance experiment, you find that the couples who had dinner in the dimmed room agreed to another date significantly more often than the couples who dined in the brighter atmosphere. A conclusion that you may consider is, all else being equal, a dim room has a positive effect on romantic attachments. However, the results would not be sufficient evidence to conclude that dim rooms cause romantic relationships. There are any number of possible reasons why a second date may not take place. The two personalities may simply clash or one of the participants could have been racist, childish, slobbish or exhibit any number of poor dating qualities. What one can take home from this hypothetical experiment is that by manipulating one variable, you may be able to conclude that it had some effect.

In the Shariff and Norenzayan experiment what was the variable? As argued, the words spirit and divine may not necessarily mean watchful gods to most people. If we return to our social psychologist thought experiment and make slight modifications, we find that it may be difficult to ascertain what impact slight changes may have on the results. For instance, if in the dim room, we added flowers, candles and romantic music and the findings saw that the dimmed room guests agreed significantly more than the bright room guests to meet again, what conclusions could be drawn? Could you select one of the variables, such as candles, and suggest that this is the cause of romantic attachments? This appears to be what Shariff and Norenzayan have done.

There was a second part to the Shariff and Norenzayan study where a third group was included, so that the second experiment included i) a control, ii) those primed with the same religious words as the earlier study and iii) those primed with secular words such as civic, jury, court and contract. The researchers conclude that the results of the second study 'showed that implicit activation of concepts related to secular moral institutions restrained selfishness as much as religious primes' (Shariff and Norenzayan 2007, p. 807). From this study, the researchers suggest that in some parts of the world, especially Scandinavia, 'these institutions have precipitated religion's decline by usurping its community-building functions' (Norenzayan 2013, p. 9). The claim is that 'these societies with atheist majorities – some of the most, cooperative, peaceful and prosperous in the world – climbed religion's ladder, and then kicked it away' (Norenzayan 2013, p. 9). It will be argued in Chapter 4 that the laws and institutions created by people were the result of a long struggle that included debate, argument and, at times, protest. Granting people greater freedoms and promoting policies that resulted in greater cooperation were made through a bottom-up process.

Cooperation and other terms

While Norenzayan and Shariff argue that the idea of God or justice stimulates cooperative behaviour because people will be frightened into being harmonious, other researchers have hypothesized that priming people with legal terms such as justice or police will cause people to be uncooperative and motivate them to be competitive and selfish (Callan et al. 2010). The researchers make the reasonable observation that the American legal system may be considered a confrontational operation, and therefore people may associate legal terms with selfishness and competition (Callan et al. 2010, p. 325). It is clear that the legal system is an adversarial process where it could be seen as a 'winner-take-all' process. The researchers point out that even the mere existence of a legal system suggests that there are those who engage in antisocial behaviour, thereby diminishing the level of trust and cooperation among people. This is supported by other studies where sanctioning and controlling the behaviour of people actually encouraged mistrust with others (Bohnet 2001). Nevertheless, these researchers hypothesize that exposure to legal terms would lead participants to behave selfishly: a hypothesis that appears to contradict the claims made by the BGP.

The results of five studies found that legal concepts were indeed associated with competitiveness and those primed with legal terms considered other participants as less trustworthy and were more competitive than those who were primed with neutral terms. The researchers conclude, 'our findings provide empirical support for the notion that people implicitly associate the law with competitiveness and that activating the law can have adverse effects on interpersonal trust and cooperation' (Callan et al. 2010, p. 327). Such contrasting results from the two sets of experiments raise serious concerns and lead to the question, how much effect does the impression of the experimenters have on the results? (Doyen 2014).

There are other studies that lead one to question the effectiveness of this type of experimentation. Consider that, in one such study, the researchers found that thinking about concepts associated with science promotes prosocial behaviour. MaKellams and Boscovich correctly note that the idea of science implies rationality, impartiality, fairness and the belief that the rational tools of science will ultimately lead to the betterment of society (MaKellams and Blascovich 2013). From this, the researchers claim that the concept of science would facilitate moral and prosocial judgments and behaviour. In another study, the same researchers found that priming people with terms such as 'hypothesis' led to an increase in prosocial behaviour. It seems that just being a science major can be associated with prosociality. In four studies, the authors found that those studying science had greater condemnation of date rape than those in other non-science disciplines. Importantly, they found moral condemnation did not correspond with religiosity or ethnicity. The authors surmise that these studies demonstrate 'the morally normative effects of thinking about science' (MaKellams and Blascovich 2013, p. 3).

From the priming literature, we find other studies where even names associated with cartoon characters apparently account for prosocial behaviour. For instance, researchers have found participants who were primed with superhero concepts were more likely to help others, although for some unknown reason, if primed with superman they were less likely to help (Nelson 2005). In the same study, the researchers found that of all the participants who were contacted ninety days after the initial experiment, those primed with superhero concepts but not superman were more likely to return to the lab again. So superhero concepts not only promote helping behaviour but also apparently, even after ninety days, are still effective.

Another experimental game used by investigators interested in human decision-making and behaviour is the PGG. In these games, all the group members have access to the public good, even if they made no contribution to it, so the reasoning assumes that each individual is motivated to freeride on the contributions made by everyone else. In a typical PGG experiment, more than two individuals form a group and they are all given the same amount of money. Each participant simultaneously decides how much money to give to the collective pot and how much to keep for themselves. The experimenters multiply the total amount of the collective pot by a number and the proceeds are distributed equally among members. If participants are motivated only by selfish needs, then they should keep everything and give nothing to the public good. Empirical evidence suggests, however, that players are not necessarily that selfish and on average contribute 40 per cent of their initial allotment. Nevertheless, experimenters use this game to see whether individuals primed with certain words will be more motivated to contribute to the public pot.

The conventional PGG begins with the experimenters giving each of four players $100.00. Each participant is asked to contribute to the common good, keeping in mind that for each dollar they contribute, an additional 40 per cent will be added by the experimenters to the common pot. To render maximum payoff for the group, each person should contribute their original allotment of $100.00 in order that the pot would amount to $560.00. Divided evenly, this would provide each person with $140.00, which is a $40.00 net gain. If one person in this scenario fails to contribute anything and everyone else puts in all their money, then there would be $420.00 in the pot. Dividing by four, each player would receive $105.00. Three of the players are richer by five dollars, but the non-cooperator walks away with $205.00. Therefore, it pays to act selfishly and not cooperate with others.

In one PGG game, researchers found individuals who were primed with words associated with cooperation were positively affected (Chaudhuri 2011). In this particular experiment, prior to the game, the participants were divided into two groups and were given ten minutes to complete a word search. In the cooperation prime, the word search was made up of five neutral words and fifteen cooperative words, such as teamwork, assist, participate, trust and support. The second group had no cooperative words, and their word search included such words as butterfly, turtle, umbrella, salad and corkscrew. The researchers found that the group primed with the words associated with cooperation contributed 11 per cent more than the non-primed group (Chaudhuri 2011).

Other researchers surmised that since there is a symbolic association between physical and moral purity, then perhaps clean smells may motivate prosocial behaviour (Liljenquist et al. 2010). After all, in a previous study the same researchers suggested that moral transgressions cause literal feelings of dirtiness (Zhong and Liljenquist 2006); therefore, perhaps cleanliness can cause 'clean behaviour' (Liljenquist et al. 2010). In this experiment, there were twenty-eight participants assigned to either a clean scented room or a baseline room (Liljenquist et al. 2010). The two rooms were identical except for the fact that the former was sprayed with citrus-scented windex. The participants played a game of trust that involved the exchange of money, and the authors of the study found that those in the clean scented room gave more money than those who were in the baseline room. They also found that the participants in the citrus-scented room were more likely to assert their willingness to volunteer at a charity, as well as to contribute money to it. From this study, the authors concluded that clean scents not only motivate people to behave virtuously, but also promoted prosocial behaviour, such as trustworthiness and being more charitable. The authors suggest that their findings are consistent with the 'broken windows' theory of crime, which is another simplistic theory without any empirical studies to establish a link between neighbourhood disorder and violent crime (see Harcourt 2001). Nevertheless, the authors suggest, 'by demonstrating that the association between morality and cleanliness is bidirectional, our research has identified an unobtrusive way – a clean scent – to curb exploitation and promote altruism' (Liljenquist et al. 2010, p. 382). Apparently this study demonstrates, 'clean scents summon virtue, helping reciprocity to prevail over greed and charity over apathy' (Liljenquist et al. 2010, p. 382).

On the basis of this review of the experimental literature, it appears that there are a number of different ways to elicit prosocial behaviour, and it is not at all clear whether some words or ideas are better than others. What is the evidentiary value of these experiments if such disparate words as spirit, divine, god, hypothesis, testing, teamwork, trust, police, jury and superhero, as well as pleasant-smelling rooms, all promote prosociality in the priming literature? It appears that the authors of all these studies assume that decisions are driven by our TASS and that our prosocial behaviour is not based on rational processes which involve choice or how we should actually treat each other. I argue that this type of evidence is the equivalent of the green shirt as discussed in the scenario at the beginning of this chapter – it is essentially useless.

Replication of experiments

As mentioned earlier, the two major general criticisms with the priming literature are the lack of replication and the problem of sample size. Since there has been no preregistered direct replication of the Shariff and Norenzayan experiment in the published research, and since the sample scale was so small, researchers have tried to replicate this experiment with slight modifications (Gomes and McCullough 2015). Since a small sample size can lead to exaggerated claims, instead of using a mere twenty-five participants in each of the cells, Gomes and McCullough included many more participants.

In the Gomes and McCullough study, there were 115 atheists and 534 theists who were tasked with unscrambling 10 five-word sentences by dropping the irrelevant word from each sentence to construct a grammatically correct four-word sentence. The participants were randomly assigned one of three conditions. The neutral prime control (NPC) group was presented with ten sentences with nonreligious words. In the standard religious prime (SRP) condition, five of the ten sentences had one religious word and four nonreligious terms, which led to religious sentences in four of the five sentences. The five other sentences were the same as the NPC. In the enhanced religious prime (ERP) condition, five of the ten sentences contained one religious word just as in the Shariff and Norenzayan study and four nonreligious terms that together completed nonreligious sentences. The other five sentences had the same five nonreligious terms used in the other two conditions. In having the SRP and ERP conditions, the authors reasoned that they would be able to ascertain whether there was a difference between priming participants with religious words embedded in religious sentences and those who were primed with the same religious words but in a nonreligious context. Additionally, since many of the participants spoke Spanish, the participants had the option of completing the game in either of the two languages.

Gomes and McCullough found that the NPC group left, on average, $4.49 to other participants. Of the NPC participants, 28 per cent left $1.00 or less, 32 per cent left exactly half the money, while 27 per cent were very generous, leaving more than $5.00. In the SRP condition, participants left, on average, $4.28 for other participants, with 34 per cent leaving $1.00 or less, 26 per cent leaving $5.00 and 27 per cent leaving more than $5.00. The participants who were in the ERP condition left, on average, $4.70 for other participants, with 29 per cent leaving $1.00 or less, 24 per cent splitting the pot and 24 per cent leaving more than $5.00. Although, as was found in the original study by Norenzayan and Shariff, there was no difference between theists and atheists, there was no evidence to suggest that religious priming increased generosity.

However, some of the findings were quite interesting. Gomes and McCullough found that women shared more than men, higher income earners shared more than lower income earners and English-speaking participants shared more than their Spanish counterparts. It would be interesting to determine whether there was any difference between the higher and lower income earners in the Norenzayan and Shariff study. If so, one may be led to suggest that the demographic composition of the participants has an effect on generosity in these types of experimental games.

Gomes and McCullough were critical of the awareness probe conducted in the original study. In the Norenzayan and Shariff study, the researchers merely asked whether the participants knew what the study was about. Gomes and McCullough also asked the participants if they recalled seeing any religious words. Out of the 600 participants, only 9 could identify the purpose of the experiment, and they were excluded from the analysis. In the SRP group, 104 participants (44 per cent) said that they knew there was a religious theme, while only 14 (7 per cent) from the ERP condition reported a religious theme. When asked whether they recalled having seen any religious words in the sentences they had unscrambled, 222 (93 per cent) in the SRP reported that they had, while 170 (87 per cent) participants in the ERP condition also remembered seeing religious terms. Here we have a study that did not replicate the original study often cited by BGP as evidence for the hypothesis that a punishing God encourages prosocial behaviour. Furthermore, a number of the participants in the ERP condition were aware of a religious theme and most participants primed in either of the religious conditions reported seeing religious words. This suggests that greater caution is required when setting up these types of experiments.

In another religious priming experiment, using a PGG rather than the DG, the researchers found that individuals contributed more money to the pot if primed with religious words (Ahmed and Hammarstedt 2011). In this experiment, the participants were divided into two groups, with one group having to unscramble religious words and the other group having to unscramble nonreligious words. Just as in the Shariff and Norenzayan study, the religious words were ambiguous and included divine, spiritual, worships, cross, paradise, god, belief and salvation. The unscrambling of sentences was similar to the other studies, as participants in the religiously primed group were tasked to take such words as '*pastry, divine, was, the and cook*' and unscramble it to read '*the pastry was divine*'. While the researchers found that being primed with religious words encouraged higher contributions, they also found that a belief in god was not correlated with contributions. That is, theists were no more likely to contribute than atheists. In fact, the finding that there is no difference in prosocial behaviour between believers and non-believers is one of the few consistent findings in the religious priming literature according to the authors of these experiments (Shariff and Norenzayan 2007; Ahmed and Salas 2008; Randolph and Seng 2007; Johnson 2010; Ahmed and Salas 2011; Lauren 2011; Gervais and Norenzayan 2012; Norenzayan 2016). I suggest that it is the only consistent finding. While they did have similar results as the Shariff and Norenzayan study, what is particularly interesting is their interpretation of the results from the believers' and non-believers' reports.

Ahmed and Hammarstedt offer two explanations to account for the results of their study. They suggest that the religious primes may have elicited thoughts of a watchful god and that this persuaded them to cooperate. This, they suggest, is in accordance with the SPH. As I have argued however, it is not immediately evident, and it is impossible to say with any confidence whether general religious primes elicit watchful, mean Gods. The second reason offered by the authors is that religious representations may activate thoughts associated with prosocial thoughts. This seems more plausible, as we find that most individuals, including some atheists, assume that there is a relationship between religion and prosociality (Galen 2012). As both Luke Galen and Nicholas Humphrey have pointed out, there is substantial evidence to suggest that there exists a general stereotype that religiosity is presumed to be associated with prosociality (Humphrey 1995; Galen 2012). Galen goes on to suggest that the majority of people assume that there is an association, and for some even a causal connection, between religiosity and morality (Galen 2012). There is some experimental evidence to suggest that simply identifying someone as religious prompts others to find them more trustworthy and moral compared to others who were classified as nonreligious (Gervais 2012). Of course, there is no evidence to suggest that this is true, but it is Galen and Humphrey's point that this is a prevalent assumption. Just because it is an assumption does not mean that it is true. Recall that throughout most of history, women were automatically excluded from human rights 'primarily because they (men) viewed them as less than fully capable of moral autonomy' (Hunt 2007, p. 28).

Furthermore, we have seen that in the priming literature, all types of words could lead to prosocial behaviour, and while most of these words are not necessarily associated with punishing gods, they all can be associated with a sense of community with other people. Rather than bringing to mind punishing gods, religious words may simply prompt people to think about their community and summon a sense of belonging. I will argue presently that the more likely explanation for the slight increase in prosocial behaviour from religious priming is due to the fact that religious primes elicit social engagement rather than punishing gods. I offer three reasons for this conclusion.

First of all, there are a number of experiments that have shown that people were more likely to act prosocially if there was a set of eyes watching them (Batson 1999). Even if the eyes were not real, the priming effect substantially persuaded people to be more honest (Bateson et al. 2006). Other experiments suggest that the amount of time that people were permitted to communicate with others prior to playing a game strongly predicted their degree of cooperation (Bohnet and Frey 1999). These experiments suggest that it is the idea of being in the presence of other people and not the idea of a mean God that may influence one's tendency to act prosocially. People interact with other people on a daily basis and are always exposed to the eyes of others. I suggest that it is much more likely that people associate eyes with other people and not an eye in the sky.

Second, the one consistent finding that is very clear with the religious priming research is the fact that both believers and non-believers are similarly affected by the primes. In other words, the level of one's religiosity has little proven effect on prosocial behaviour. This suggests that it is the social aspect of the primes rather than a metaphysical belief in punishing gods that is activated in the minds of the participants. After all, why would an atheist be influenced by a punishing God?

It makes much more sense to reason that the common aspect of the words for both believers and non-believers used in the priming experiments is that the words are associated with a community of people. After all, there is evidence to suggest that atheists who happened to be involved in Church life are much more likely to volunteer in a soup kitchen than a believer who prays alone (Putnam and Campbell 2010, p. 472). It is not inconsistent to believe that a devout atheist would volunteer his time with helping out at the local church. I suggest that atheists are much more likely to relate religious words with community involvement than with the idea of being watched by an angry God. The BGP offer no evidence to counter the claim that religious individuals are influenced by community engagement rather than mean Gods. Consider the findings of Robert Putnam and David Campbell, who found that while religiously based social networks accounted for most of the apparent effects of church attendance, they also found that 'religiosity becomes entirely insignificant as a predictor of virtually all measures of good neighbourliness' (Putnam and Campbell 2010, p. 472). This is consistent with my claim that many religious people are good neighbours, but the idea of God is not the reason why they are good.

Third, researchers suggest that religious, more so than nonreligious, individuals are motivated to maintain a good reputation and that they are very concerned about reputation management. Indeed there are a number of studies suggesting that religiosity predicts prosocial behaviour when others have a positive image of the individual (Batson 1989). Other studies suggest that religious individuals tend to focus on appearing to be helpful, rather than participating in actual helpful behaviour (Batson 1990). Furthermore, 'studies repeatedly indicate that the association between conventional religiosity and prosociality occurs primarily when a reputation-related egoistic motivation has been activated' (Norenzayan and Shariff 2008, p. 59). If the reasoning is that people resist moral transgressions because of a fear of a supernatural punisher, then why do religious individuals have a heightened concern for their reputation? Rather than an internalization of a mighty monitoring God to promote good behaviour, it is more likely that the good behaviour is prompted by the thoughts of other people.

Mean or benevolent gods?

Since the hypothesis put forward by the BGP rests on the idea that mean Gods make good people, the researchers make a concerted attempt to demonstrate that the idea of hell is a stronger motivator than heaven. Take, for instance, the claim that 'apparently, a punishing God keeps people in line' (Norenzayan 2013, p. 45). Norenzayan cites a study by Amber DeBono, Azim Shariff and Mark Muraven to support this claim. Norenzayan explains that in the experiment conducted by DeBono and colleagues, Christian participants were randomly divided into two groups. One group formed the forgiving God category, where the participants were given a Biblical excerpt from James 3:17 to read and then asked to write about a Forgiving God. The second group read from Deuteronomy 29:20 and were asked to write about the punishing behaviour of God. After the priming condition, all were asked to take part in a study where they were invited to solve anagrams. They were told to pay themselves one dollar for each anagram they solved correctly. The anagrams were rigged so that some were unsolvable, yet most of the participants paid themselves anyway. According to Norenzayan, the researchers found that the forgiving God group overpaid themselves significantly more than the punishing God group. From this, Norenzayan concludes that a punishing God is more effective than a forgiving God at keeping people honest (Norenzayan 2013, p. 45). What is interesting, however, is the part of the study that Norenzayan left out of his book (Norenzayan 2013).

According to Shariff and Rhemtulla, there was a third condition in the DeBono unpublished study (Shariff and Rhemtulla 2012). The third group consisted of individuals who were primed with neither a punishing nor a forgiving god. The researchers of the original paper found no statistical difference between the neutral group and the punishing God group in overpayment to themselves. This information is very significant for two reasons. First, one would think that since the participants were all Christians, then those who were primed with a punishing God would have cheated less, since that would be in accordance with the SPH. Second and more importantly, we have seen that in other priming experiments, the religious words such as divine and spiritual were ambiguous, yet the researchers still suggested that the participants' behaviour was affected by the belief in a punishing God. In the DeBono study, we have the prime that categorically emphasizes the punishing aspect of God, rather than general religious terms, yet there is no statistical difference between the punishing God group and the neutral group. Rather than supporting the SPH, the DeBono experiment raises serious doubt about the role of a punishing God for prosocial behaviour.

While it has been argued that one must be very cautious when drawing conclusions from the results of the number of priming experiments on religiosity and prosociality, researchers nevertheless make bold statements about the results of these priming studies. For instance, Shariff and Rhemtulla say, 'numerous lab studies have established direct *causal* effects for religious beliefs on both prosocial and antisocial behaviour' (Shariff and Rhemtulla 2012, p. 3 *my italics*). They say that much research supports the conclusion that the fear of supernatural punishment deters transgressive behaviour. Having suggested this causal claim, these researchers go on to refer to the DeBono study. While Shariff and Rhemtulla include all three groups in their use of the study, they particularly focus on the difference between a benevolent god and a punishing god. They reason that if there is a difference at the individual level, then we should be able to predict transgressive actions, such as criminal behaviour, at the societal level. Their investigation into international crime rates suggests that an increase in the belief in heaven predicted higher crime rates, while belief in hell predicted lower crime rates. They surmise that although these findings may seem unrelated to the lab studies, they make the claim that 'parsimony suggests that both are, at least to some degree, a reflection of the same underlying *causal* story' (Shariff and Rhemtulla 2012, p. 3 *my italics*). When it comes to crime prevention and the belief in a punishing God, I argue that the BGP are skating on very thin ice.

The Shariff and Rhemtulla study focused on the relationship between a belief in heaven and crime and a belief in hell, and national crime rates. The study neglected to consider the relationship between no belief in either heaven or hell and crime rates. From this research, one should not make the claim that a belief in hell causes a reduction in crime. Yet Norenzayan, citing this study, says, 'all else being equal, countries with high percentage of believers in God had lower crime rates' (Norenzayan 2013, p. 46). Surely, these researchers can't be serious in suggesting that the countries with the majority of people believing in punishing Gods have the safest societies. It seems misleading to make this claim without factoring in the relationship on crime rates and no belief in God.

According to the Institute for Economics and Global Peace Index 2016, the most peaceful countries in the world include Iceland, Denmark, Austria, New Zealand, Portugal, Czech Republic, Switzerland, Canada, Japan and Slovenia. The least peaceful are Iraq, Afghanistan, Somalia, Yemen, Central African Republic, Ukraine, Sudan and Libya. One can't help but notice that the most peaceful countries are the least religious and have the institutions and structures that support an open society, while the least peaceful make up the countries that have the most believers in God and untrustworthy institutions.

To support the claim that the temptation to cheat cannot be overcome by rewards nearly as well as it can be overcome by a threat of severe punishment, the BGP offer as evidence another study by Shariff and Norenzayan. In this study titled 'Mean Gods Make Good People', the researchers had 67 undergraduates solve math problems at a computer terminal (Shariff and Norenzayan 2011). The participants were cautioned that there was a glitch in the computer and sometimes the answer would appear. The participants were told that if this happened, they need only to strike the spacebar as fast as they could and this would alleviate the problem. The researchers assessed the number of items, out of 20, on which participants failed to press the spacebar in order to measure the level of cheating. Following the cheating task, participants were each given a questionnaire with a number of traits describing God. The traits included seven positive qualities such as a forgiving, loving and compassionate, while negative views included vengefulness, harshness and anger. They were asked how much each of these traits applied to God, and if they were nonbelievers, what they believed others in their culture thought about God. What the researchers found was that a God, described in negative terms, was associated with lower levels of cheating. In this study, just as in previous ones, there was no difference in cheating between believers and atheists. The authors said that cheating was uncorrelated with intrinsic religiosity. What was interesting was that women cheated more than men.

The authors did a second study that included more than just undergraduates. The results were the same in that punishing Gods predicted less cheating and self-described believers were no more or less likely to cheat than atheists. Unfortunately, the authors didn't include a third variable to see whether a view on institutional punishment had any effect on cheating behaviour. For instance, would those who have a tough-on-crime attitude, calling for harsher and longer sentences, be less likely to cheat? What is extremely important here is that the authors of these studies encourage people to take great care in these types of experiments as 'the link between views of God and cheating behaviour revealed by the current data is a correlational finding and therefore should be interpreted with caution' (Shariff and Norenzayan 2011, p. 94). The researchers go on to suggest that making such a provocative claim that a comforting God rather than a punishing God is related to greater levels of cheating 'certainly requires more evidence before it can be made with any confidence' (Shariff and Norenzayan 2011, p. 92). The problem is that there has not been any credible evidence to support the claim made by the BGP. In fact, there are some authors who suggest the opposite of what the BGP claim and suggest that they have experiments to back up their assertion.

While the BGP propose that an authoritative, punishing God facilitates prosocial behaviour because people expect external rewards and punishments for their behaviour, others have suggested a benevolent God encourages prosocial behaviour because this type of belief fosters an internalization of a benevolent self-identity (Johnson, K. 2015). There is research to suggest that rather than promoting prosocial behaviour, a belief in a punishing, authoritative God is associated with depression and neuroticism (Rosmarin 2009). There is some evidence to support the claim that the majority of Christians represent God as benevolent (Johnson, K. 2014). If that is the case, then even if a belief in a mean God has any effect at all, it will be slim. Other researchers speculate that if Christians see God as benevolent and express their altruistic values by helping others, then representations of a benevolent God would be positively associated with a benevolent self-identity, which would be greatly related with helping others (Johnson, K. 2015).

As previously mentioned, the majority of the experiments in the priming literature by the BGP involved subjects who were primed with general religious terms rather than specific ideas about the nature of God. In contrast, the researchers conducting a different study were interested in determining whether the belief in a gracious, loving and merciful God would have a positive effect on prosociality. They focused on secular volunteering, defined as acts of helping others who are outside one's family or religious group (Johnson, K. 2015). Using Christian undergraduates, the researchers found that while an authoritarian God had a negative effect on secular volunteering, a benevolent God is more 'strongly and positively associated with both previous secular volunteer experience and intentions to help in the future' (Johnson K. 2015, p. 7). They conclude that the representation of a benevolent God was a positive predictor of a benevolent self-identity and a belief that they had a moral obligation to help. This obligation to help came from a moral duty to help others and was not necessarily a religious duty. This is consistent with other studies that suggest people volunteer as an expression of their prosocial values and not because of a belief in God (Okun 2014). Where a representation of a benevolent God was associated with a belief to help others, an authoritarian God representation did not have such an effect (Johnson, K. 2015, p. 9). These studies clearly counter the claim that mean gods make good people.

Punishment or fairness?

The argument that a watchful God is either a biological or cultural adaptation that evolved to sustain human cooperation lies with the assumption that fear of a punishing God deters cheating or uncooperative behaviour. In other words, the core idea is the deterrence of antisocial behaviour. Yet it could reasonably be argued that people are more concerned with fairness than punishment. Some have suggested that humans have what may be called a fairness instinct rather than a punishment instinct (Sun 2013). In Chapter 4, I will look at what we mean by fairness and whether it is innate or not, but for the moment, let's consider some ideas about fairness. When a tragic incident occurs, people typically speak on terms of fairness rather than referring to group safety or the prevention of repeat occurrences. Consider a failure to remain at the scene of a traffic accident where three children are killed by an impaired driver with an alcohol intake of twice the legal limit. At trial, defence, in arguing for a lighter sentence, would focus on the fact that the accused had turned himself in, is extremely remorseful and is progressing well in an alcohol abuse programme. This suggests that the accused has taken full responsibility, is unlikely to commit further crimes and is no longer a threat to the community. If people strictly saw punishment as a means of deterring antisocial behaviour as the BGP suggest, then the public would be receptive to such an argument offered by the defence, as there is no concern about detecting the perpetrator or the likelihood of a repeat occurrence. Generally this is not what one sees in the media or hears in public discussion. What is stressed is the unfairness of a situation where the accused receives a light sentence in relation to the essential life sentences handed to the children. People become angry due to the unfairness of the situation and don't generally take into consideration deterrence.

Crime and punishment

It has been argued that the evidence to support the claim that a fear of a punishing God motivates people to be good and to avoid antisocial behaviour is largely ambiguous and open to all sorts of interpretations. Thus far, it has been demonstrated that the most-prized priming evidence disclosed by the BGP is open to doubt. While the experimental evidence from the laboratory is suspect, the evidence proposed by the BGP concerning punishment and crime prevention, or more appropriately, crime reduction, is simply incredulous. Consider the suggestion that secular deterrents are 'pathetic compared to an eternity in hell' and that in most Western countries 'people are dissuaded by crimes only by prison and fines', which 'may be more humane, but is less of a deterrent' (Johnson 2016, p. 228). These allegations raise two separate but not unrelated issues that will be addressed presently.

The first issue deals with the idea of secular deterrents being pathetic. Pathetic, according to Johnson, means that the punishment is not harsh enough. Yet historically, humans have devised all types of cruel punishments, such as the British custom of being hanged, drawn and quartered. Breaking on the Wheel was designed so that death would not proceed too quickly. The executioner would tie the condemned man to 'an X-shaped cross and systematically crush the bones in his forearms, legs, thighs and arms by striking each one with two sharp blows' (Hunt 2007, p. 72). This would be followed by the executioner dislocating the vertebrae of the prisoner's neck before fastening his 'limbs bent excruciatingly backward to a carriage wheel on top of a ten-foot pole' (Hunt 2007, p. 72). There the body would remain for all the people in the town to see. In the United States, 'women were taken from a huge cage, in which they were dragged on wheels from prison, and tied to the post with bare backs, on which thirty or forty lashes were bestowed amid the screams of the culprits and the uproar of the mob' (Hunt 2007, p. 78). Slaves, on the other hand, caught for arson would have their right hand cut off, hanged, head cut off and the dismembered body parts displayed for the public. This is only a taste of the human capacity to inflict cruel and unusual punishment.

The issue is not whether we can dish out crueller punishments in order to allegedly deter people from committing crimes, but more importantly how, as a society, did we manage to stop that behaviour? Did it come from above or from below? I suggest that it was accomplished through a slow campaign by individuals bringing to the public forum a number of persuasive reasons why this type of cruel behaviour must be stopped. This campaign planted the seed for further progress on human rights. In Chapter 4, I will argue that the recognition of human rights was the essential ingredient for 'thick cooperation'. Rather than suggesting that punishment by jail and fines is pathetic, I will argue that the drive for reducing harsh punishments to a more reasonable system of justice and fines was closely affiliated with the rights movement, paving the way for the idea that every human being is actually worthy of rights.

Furthermore, John Tooby and colleagues conducted a number of simulations to see whether cooperation within a group could emerge without punishment (Tooby et al. 2015). They found that even without punishment, 60 per cent of the time everyone in the group cooperated. They found when there were adequate gains from cooperating, then 'cooperation evolves, sometimes substantially so' (Tooby et al., p. 9). They also found that it did not take much punishment to encourage cooperation, and this was especially evident when groups were large. They concluded that even rare punishment could induce much cooperation (Tooby et al., p. 14). This brings us to the second issue dealing with the question of whether harsh punishment, in the real world, is actually effective in reducing crime.

There are studies suggesting that prisoners who were held in particularly severe institutions were more likely to reoffend than those held in less-harsh prisons (Drago 2011). Furthermore, while noncustodial sanctions are, more often than not, more effective than jail terms, it is unclear why some of these sanctions are better than others (Villettez 2015). It may very well be that the determining factor for encouraging subjects to respond better to some sanctions over others is due to the perception that the sanctions seemed to be fair. Further research is needed in this area.

There is considerable debate on whether custodial or noncustodial sanctions have a greater effect on whether offenders will reoffend or not, and the evidence to date is inconclusive. By 'custodial sanctions', it is meant that an offender's freedom of movement is deprived, while 'non-custodial sanctions' refer to such things as fines without a deprivation of liberty. Logically then, prison sentences would be harsher than fines. In a recent meta-analysis on the impact of imprisonment, researchers found that imprisonment is not only less effective than noncustodial alternatives but also, according to a majority of quasi-experimental studies, less damaging (Villettaz 2015). This is consistent with other reviews, where it was found that incarceration is followed by higher rates of recidivism than noncustodial sanctions (Smith 2002). In a recent study on crime in the United States, researchers found that incarceration had little to do with the decline in crime and that continuing to incarcerate people at a high rate will have no effect on reducing crime (Roeder et al. 2015). In another study, researchers found that security cannot be achieved through threats of punishment or harsher sentences; rather, the most effective way of decreasing crime is to increase the number of police officers (Nagin 2012). In one study on the effects of the three strikes law in California, the researchers found that while the legislation had a deterrent effect on property crime, there was no such effect on violent crime (Kelly and Datta 2009). In another study on the long-term effect of the California legislation, the researchers found that the added deterrent of the third strike was considerably less than the second strike, questioning the justification for such harsh punishment (Datta 2017). All these studies suggest that people will more likely reoffend after a prison sentence and the harsh punishment is not a deterrent. Yet, it is difficult to make broad conclusions, as there are just too many conflating factors; therefore, one can only infer that there is not enough evidence to suggest that a threat of harsh punishment is an effective method to prevent crime.

The war on drugs is a prime example of the irrationality of the claim that the threat of harsh punishment leads to prosocial behaviour. Consider that in the United States there are approximately 400,000 inmates for drug offences. This figure is higher than the number of people incarcerated in England, France, Germany and Japan for all crimes combined. Yet drug trafficking and possession and the violent crimes associated with drug trafficking continue in the United States (Gray 2001, p. 30). Doctors who deal with people affected by drug addiction sensibly argue that the drug situation should be viewed as a medical problem rather than a crime issue (Mate 2008). These arguments from doctors make sense since research suggests sentencing to prison for simple possession had significantly higher rates of recidivism (Spohn 2002). In another study involving first- and second-time offenders convicted of impaired driving, researchers found that the most effective sanction was a licence suspension and a rehabilitative alcohol programme, not jail (Tashima 1989). All substance-abuse-related crimes are very difficult to assess as there are so many variables involved that contribute to the situation of the offender. Therefore, harsh prison sentences, or the threat of retribution for some offences, are much more likely to be detrimental. Threatening people with either supernatural or harsh secular punishment is not going to relieve the problems associated with the current Fetanyl problem in North America. What we can conclude, however, is that the BGP assertion that the threat of punishment is a powerful force for deterring antisocial behaviour is questionable to say the least.

There have been other methods used by the state to try to deter crime. One is the implementation of minimum sentences. In Canada, an individual faces a minimum sentence if convicted of one of a number of serious criminal offences. In a recent Supreme Court of Canada case, where the accused was originally convicted of firearm offences, the defence appealed to the Court of Appeal arguing that a three-year minimum sentence offended sections 12 and 15 of the Charter of Rights and Freedoms (R vs Nur 2015, SCC 15). The Appeal Court ruled that even though the mandatory minimum terms of imprisonment are grossly disproportionate sentences to reasonable hypothetical cases and therefore violate the Charter, in the *Nur* case, the sentence imposed was appropriate. The Supreme Court upheld that decision, but what is relevant to my argument here are the comments made by the Supreme Court Justices. Quoting from the Sentencing Reform: A Canadian Approach – Report of the Canada Sentencing Commission, the Justices said, 'it is extremely doubtful that an exemplary sentence imposed in a particular case can have a perceptible effect in deterring potential offenders' (R vs Nur 2015, SCC 15 paragraph 113). They went on to say, 'law-based, punitive measures alone cannot produce large sustained reductions in the magnitude of the problem' (R vs Nur 2015, SCC 15 paragraph 114). They also referred to research by University of Toronto criminologists concluding: 'empirical evidence suggests that mandatory minimum sentences do not, in fact, deter crimes' (R vs Nur 2015, SCC 15 paragraph 114). From the wisdom of the Supreme Court Justices, we find that the allegation that threats of harsh punishment or the imposition of minimum sentences has little effect in reducing crime again raises serious doubts about the BGP claims.

Another misconception generated by the BGP involves the idea of punishment based on the retributive function, known as 'Just Desserts'. Dominic Johnson, in *God Is Watching You*, identifies a number of basic cognitive underpinnings that form the foundation for religious thought. He lists cause-and-effect reasoning, mind–body dualism, the HADD and a belief in a just world. I dealt with the first three in the previous chapter and will focus attention here on what Johnson claims is a key characteristic of religious thinking, that is, the just world hypothesis. Johnson quotes ethicists Clare Andre and Manuel Velasquez as follows: 'people have a strong desire or need to believe that the world is an orderly, predictable and just place, where people get what they deserve' (Johnson 2016, p. 126). Perhaps some people do in fact believe this, and Johnson does offer a number of studies to support his point; however, he goes on to quote the authors' claim that, 'when we encounter evidence suggesting that the world is not just, we quickly act to restore justice by helping the victim or we persuade ourselves that no injustice has occurred. We either lend assistance or we decide that the rape victim must have asked for it' (Johnson 2016, p. 126–127). Now this is problematic.

It is quite clear that people are concerned with injustice, but we need to determine whether it is related to an orderly world where people get what they deserve or is it related to a consideration of justice? After all, studies with children point to their concern for fairness and justice, and they typically come to decisions by reasoning about how people are treated (Tomasello 2014). I will have much more to say about fairness, justice and reasoning in the following chapters, but for now, I suggest that people typically do have a concern for justice and fairness, but the particular culture strongly influences who is to be considered worthy of fair treatment or not. The fact that some people could actually consider a rape victim as 'asking for it' has more to do with ignorance and a misconception about sexual assault victims than contemplating an orderly world. Presumably, the idea that a rape victim 'asks for it' is related to the behaviour of the victim, behaviour such as consuming alcohol or placing oneself in a volatile situation.

Yet, if we consider a scenario where a man after consuming alcohol leaves the local tavern at closing time, cuts through a dark laneway on his way home, only to be robbed and severely beaten by a group of thugs, it is very unlikely that the behaviour of the victim would be taken into account. If this situation was compared with another robbery, where the male victim was robbed in the middle of the day, at a busy intersection, in a downtown city, most likely the two robbery victims would be treated in a similar fashion. No one would suggest that the first robbery victim had somehow 'asked for it'. The reason why an increasing number of people no longer have these misconceptions about sexual assault victims is due to small clusters of individuals who lobbied against these mistaken views. Advocates have managed to convince others that sexual assault myths are simply ignorance masquerading as knowledge. This is the way people have managed to widen the cooperative circle and dispel myths: by reasoning and arguing about the issues.

I am not suggesting that we have managed to dispel the myths of sexual assault by any stretch of the imagination. Consider the Supreme Court of Canada ruling, where Madame Justice Heureux-Dube chastised a foolish remark by a Judge on the Alberta Court of Appeal by saying 'these comments made by an appellant Judge help reinforce the myth that under such circumstances either the complainant is less worthy of belief, she invited the sexual assault or her sexual experience signals probable consent to further sexual activity' (Ewanchuk 1999, SCR 330, p. 89). The Supreme Court Justice rightly states, 'myths of rape include the view that women fantasise about being raped ... that women often deserve to be raped on account of their conduct, dress and demeanour' (Ewanchuk 1999, SCR 330, p. 82). When hearing arguments accounting for a 'just world', we should be very cautious about any accompanying myths that follow from the hypothesis.

Furthermore, the BGP make such unsubstantiated claims as 'whatever earthly punishments may be understood to lie in store for us, people tempted to do wrong often are (or become) concerned that they will face supernatural punishment as well' (Johnson 2016 p. 192). It would be very difficult to document whether people often think about big punishing gods prior to committing crimes, let alone having God on their minds at all. In fact, there is some evidence to suggest that many offenders are overconfident in their belief that they will escape detection and rarely have concerns about future consequences (Robinson and Darley 2004). Offenders don't always contemplate what others are thinking or what punishment is in store for them, let alone caring about the possibility of supernatural punishment in the distant future or in a life to come. Offenders may actually use a cost/benefit analysis to determine their choice of crime to commit. For instance, a person convicted of the offence of trafficking marijuana in Canada would most likely receive a light sentence, quite possibly avoiding incarceration, while a conviction for trafficking in cocaine or heroin will most likely incur a jail term. A person may choose to engage in dealing in substances which could result in a lesser or no jail time, even though it may not be as lucrative. They may opt to commit frauds rather than robberies because they would more likely receive a noncustodial sentence for the former. Additionally, I have met many people accused of crime who saw fines or jail as simply the price of doing business, a sort of occupation tax, and they weighed the associated costs and benefits of their chosen paths. If they happened to be apprehended, they hoped their lawyer would get them the best deal possible.[1]

There are number of reasons why people commit crimes, so it seems overly simplistic to propose that the threat from nasty supernatural agents actually prevents crime. For instance, a person may be raised in a family where the grandfather, father and brothers were career criminals, so it may be very difficult to escape from the 'family business'. Others may have been dealt a bad hand in life. For instance, many young prostitutes have suffered sexual abuse at the hands of their guardians. To escape such abuse, they ran away from home, and this eventually led to the use of illicit drugs. To pay for their drug addiction, they had to sell their bodies to make money. These victims of crime certainly do not need to be threatened with fiery hell nor, I would suggest, would such a threat have any effect at all in preventing criminal behaviour. What they need is human intervention, compassion and assistance. In addition to those who fell from grace due to unpredictable circumstances, there are others who are simply self-centred and uncaring, such as David Cash, who was introduced at the beginning of this book. There are others, such as Cash's narcissistic sociopath friend Jeremy Strohmeyer, who had little concern for anyone and would readily assault people regardless of any threat of punishment. Fortunately, there are not too many people like either of them, and the majority of people obey the law and do not need to be threatened with severe punishment (Keihl 2014). The claim that a fear of God prevents crime is overly simplistic and not supported by the evidence.

It is unlikely that humans will ever prevent crime, but there are a number of effective solutions that have proven to be powerful methods in reducing crime. For instance, trustworthy law enforcement based on a legitimate rule of law with a fair fact-finding judiciary and moderate punishment is the most effective way to reduce crime. Harsher penalties may elicit votes for politicians who push the emotional buttons of certain segments of society, but as a legitimate method for reducing crime, the hypothesis does not survive a preliminary hearing.

Constant surveillance

One of the main arguments from the BGP is the notion that the gods closely monitor human activity. For instance, some scholars argue that 'the belief that one's actions are constantly and inescapably being observed by a divine being may be a strong stimulus and reminder to be aware of one's actions' (Baumeister 2010, p. 76). Others have suggested that 'as groups expand, Big Gods emerge, who demand sincere commitment, micromanage humans round the clock, punish transgressions and reward good behaviour' (Norenzayan 2013, p. 131). The point stressed by these researchers is that humans are consciously and subconsciously aware of this unrelenting Orwellian surveillance by the gods. Yet some of the evidence offered raise serious doubts as to the effectiveness of such surveillance. For instance, one study by Benjamin Edelman is cited as evidence for the Big God Hypothesis, but it seems to me that the evidence in its support is rather weak (Norenzayan 2013, p. 36). When Edelman compared the frequency with which people in the United States visited Internet pornography sites from highly religious states with people in not so religious states, he found no difference. What he did find however was that those from the religious states refrained from porn consumption on Sundays, but made up for it during the rest of the week.

The question is, if gods are 'constantly monitoring people 24/7', then why do religious people only behave, as if being watched, on Sundays? The answer from Norenzayan is that religion is more in the situation than in the person (Norenzayan 2013, p. 39). A more persuasive reason as to why religious people from religious states may be dissuaded from watching porn on Sundays has to do with the social climate. Ostensibly, the more religious people would have attended church on Sundays and therefore would have interacted with other people. Perhaps the social atmosphere and a sense of community dissuade porn consumption. Maybe the further we are away from that social atmosphere, the more willing people are to surf porn sites. What we can say with confidence is that we simply just don't know why people behave in such a fashion, and we can be certain that the Edelman study provides evidence to counter the efficacy of constant surveillance by God to uphold moral behaviour and thereby lends no support for the SPH.

Prior to concluding this chapter, I will provide additional examples to demonstrate that the SPH is not quite up to the hype. Dominic Johnson cites the 1969 Montreal police strike, where the cops went home instead of patrolling the streets, to support his hypothesis. 'Within minutes of the strike beginning, there was a spate of burglaries, bank robberies, murders, rioting, and looting', argues Johnson (Johnson 2016, p. 229). For Johnson, this shows that the 'police may be necessary but they are not infallible'. Of course, he says, 'God does not go on strike' (Johnson 2016, p. 229). If a fear of supernatural punishment dissuades people from crime, then what happened in Montreal? What we learn from the Montreal situation is that the police are certainly necessary, but we have to look at the overall picture before concluding that there was total carnage in the streets of Montreal.

The police in Montreal went on strike due to a wage dispute as they wanted to be paid the same as their brothers and sisters in Toronto, and it is true that there were a total of six bank robberies in Montreal that day, which is more than typical. However, the first bank robbery did not occur within minutes of the walkout (Pinker 2011). There were two murders during this time, but one of the people murdered was a Quebec Provincial Police officer who was shot during a riot. This riot was the result of a protest by taxi drivers who were objecting to the fact that a rival company had exclusive rights to customers at the airport. Union officials took advantage of the fact that the police were on strike, and this led to the violence. It was not as if neighbours were raping neighbours and people were quivering in their basements; rather, a very few number of individuals took advantage of the circumstances to capitalize on their self-centred motives. Most people are generally law-abiding because of the institutions that are in place, and what is critically important is the fact that these organizations are legitimate in the eyes of the people. I will pursue this logic in the chapters to come.

Furthermore, the idea that there is a need for constant divine surveillance doesn't make much sense when it comes to such companies as eBay. Believing that people are essentially good, Pierre Omydiar launched the idea in the mid-1990s. Today the company has more than 200 million users, and the people buy and sell their goods without knowing each other or whether they are fearing the same God. Remarkably less than 1 per cent of the eBay transactions are unsuccessful (Pfaff 2015, p. 182). When there is a problem, users are encouraged to communicate with each other and work it out. That is, they are advised to reason through their problem and try to come up with a solution. This way of resolving the problem results in a 50 per cent success rate on those very small numbers of problems (Pfaff 2015, p. 182). It is not just companies such as eBay,

Kijiji and Craigslist that puts a damper on the BGP argument but also consider such successful companies as HomeExchange. The company boasts that they have 65,000 members with hundreds of thousands of successful exchanges. People all over the world open up their homes, exchange cars and take care of each other's pets without a need for God watching them or their valuables.

Summary

Let's summarize the evidence looked at in this chapter that purportedly supports the BGP contentions. We found that in some studies, subjects primed with 'religious' concepts, but not necessarily God terms, were prone to be more generous. Other studies, however, did not reach the same conclusion. When primed with justice terms, we found that some participants were generous, while others were not, and in some experiments the majority of people primed with justice terms were more competitive rather than cooperative. Other studies suggested that science majors condemned date rape more than students from the humanities and those participants primed with science terms encouraged prosocial behaviour. Interestingly, moral condemnation did not correspond to religiosity. Even people primed with superheroes were more generous, unless for some reason the prime was superman. In other studies, people primed with words such as 'teamwork' tended to be more generous, while in other experiments those primed with 'hypothesis' were more prosocial. In other studies, researchers found that those primed with a punishing God cheated less, while other studies showed that people primed with a benevolent God were the nicer people as they were more willing to help others and those primed with a punishing God were not quite so eager to lend a helping hand, particularly if the recipient of the benevolent act was an outsider. Judging by the disparate results and interpretations from the priming literature on prosociality, it would not be surprising at all if researchers found that priming participants with such random words as 'fire hydrant', 'chicken vindaloo', 'picnic tables', 'slinky's' and 'sunsets' would also lead to a correlation between those words and prosociality. Furthermore, in nearly every priming experiment to date, there is no difference in prosociality or generosity between religious and nonreligious people. This questions the soundness of the BGP claim and suggests that there must be other reasons for why people are prosocial. Finally, although the BGP thesis rests with the idea that mean gods make good people, in the criminology literature we found no evidence that harsh punishment or the threat of such punishment is effective in the real world. In contrast, some studies suggest that harsh punishment is actually detrimental. The only definitive conclusion that can be made from the experimental research discussed in this chapter is that no conclusion can be made as to the efficacy of the idea of God in promoting prosocial behaviour.

If it isn't the fear of God that keeps us in line and encourages us to cooperate, then what is it? In the following chapter, I will argue that humans are not the Hobbesian creatures portrayed by the BGP. Rather humans have what Patricia Churchland calls a 'neural platform' for cooperative behaviour (Churchland 2011). The platform, I will argue, is biological, and in order to understand how we managed to scale up from a hunting and gathering society consisting of small groups of mostly related individuals to societies of people accepting strangers requires an evolutionary explanation. In the next chapter I will argue that the explanation for human cooperation begins with the irresistible drive to engage with others.

Note

1   I spent more than thirty years working in the Canadian Justice System.

3

Family matters

Imagine receiving a phone call from your brother and he expresses some unsettling news. He has just returned from the doctor's office after being told that he has hepatitis C, and without a liver transplant, he will soon die. He hesitantly asks whether you would be willing to undergo surgery, so that the doctors can remove part of your liver and transplant it to him. You would be hospitalized for a while, and it would take over six months for your liver to grow back to about 90 per cent of its original size. This would cause considerable discomfort, and there is always the possibility that things could end gravely and you may die. You are certainly paying a cost to benefit someone else. Yet, without hesitation you agree to the procedure and make arrangements to see whether you are a suitable candidate. Why do we do such things for our kin? Throughout the world, there are numerous people on waiting lists for similar operations, yet they will inevitably die due to the lack of individuals willing to do for others what people will naturally do for 'their own'.

The idea that kin is key to understanding human relationships has a long philosophical history. For instance, in *Nicomachean Ethics*, Aristotle said that friendships and justice seem to be concerned with the same things. In every community, we find certain types of friendships, as well as certain types of justice. However, he reasoned, justice is different, depending on who was involved. Aristotle said, 'it is more dreadful to defraud a comrade than a fellow citizen, to fail to aid a brother than a stranger or to hit one's father than anyone else' (Crisp 2014, p. 154). The friendship topping the list for Aristotle is the natural one, between a parent and child, and this special relationship not only concerns human beings but also is found with birds and other animals as well. Today, we wouldn't necessarily call the relationship between a parent and child a friendship; however, Aristotle was quite correct in noting the significance of this relationship. In the previous two chapters, I looked at the evidence offered by researchers to support the claim that it was prosocial religions, with their belief in a punishing God who watched and intervened in human affairs, that paved the way for cooperation in large groups of anonymous strangers. The foundation of this argument rested on the belief that a fear of God would promote prosociality and curb people from deviating from group norms. Furthermore, the BGP claimed that groups composed of such people would outcompete groups of people who were not inclined to believe in supernatural agents. By looking at the evidence offered by the BGP, I suggested that they failed to provide a *prima facie* case. To understand why humans are getting better at getting along, I submit that we must dig much deeper into our past and consider the evolution of the mammalian brain.

In this chapter, I argue that the mammalian brain evolved to seek mates and to take care of kin in much the same way that we take care of ourselves (Churchland 2013). It is reasonable to suggest that there would not be any cooperation at all, unless there were biological mechanisms in place to provide a foundation for prosocial behaviour. I will show that the human capacity for cooperativeness and helping behaviour is built on a self-interest platform. The first part of this chapter will look at the neural and chemical underpinnings that drive us to search for mates and bond with our children. While all mammals have minds for the capacity for helping behaviour for kin, it is essentially only humans who extend cooperation to include many strangers. While the BGP suggest that there is a need for a fear of God to bind groups, I argue that our brains are already wired to motivate us for 'thin cooperation'. In fact, humans are dependent on each other for survival. Evidence suggests that humans, as well as all primates, have little difficulty in forming coalitions. We have 'tribal social instincts' that lead us to cooperate naturally with our kin and in-group, but these same instincts make it difficult to promote cooperation in larger groups (Richerson and Boyd 2005). By looking at how the hominin brain evolved to cooperate with others in a restricted way, we will find that the game changer that led humans to have the capacity to overcome our biological limitations was the evolutionary pressures for language.

Kin selection

William Hamilton provided the underlying process in the evolution of cooperation in his theory of kin selection. He reasoned that a 'gene causing its possessor to give parental care will then leave more replica genes in the next generation than an allele having the opposite tendency' (Hamilton 1964, p. 1). He went on to note that the parent/offspring relationship is similar to the full-sibling relationship. He concluded that if an individual carries a certain gene, it then stands to reason that the chance of a sibling carrying that gene is 50 per cent. For Hamilton, an individual advances its own genetic future by assisting kin who carry copies of its genes. So parents will make sacrifices on behalf of their children, brothers will do the same for their sisters and so on. This makes sense, as otherwise self-sacrifice would be a self-defeating genetic strategy. However, if we consider an animal's own reproductive success, as well as the indirect effects on the fitness of kin, then this will lead to helping others or prosocial behaviour in the species. As argued in an earlier chapter, fitness for biologists means the individual's ability to pass on genes to the next generation (Dawkins 1989). Hamilton called this inclusive fitness and suggested helping behaviour can be expressed in mathematical terms. According to Hamilton's rule, helping behaviour should evolve when the cost (c) to the helper is less than the benefits (b) derived from helping an individual who is related (r). For Hamilton, we should be able to predict the degree of cooperation by using the formula $r > c/b$.

Greedy reductionism

The suggestion that the investigation into human cooperation should begin at the level of brain chemistry may arouse fears of reductionism; however, such uneasiness should be put to rest once the meaning of the word is untangled. Daniel Dennett distinguishes between what he considers a good form of reductionism, like evolution by natural selection, and its weaker cousin, 'greedy reductionism'. In the context of Darwinian theory, the distinction is best understood by suggesting that 'greedy reductionists think that everything can be explained without cranes; good reductionists think that everything can be explained without skyhooks' (Dennett 1995, p. 82). By no means is it being suggested here that all evolved behavioural patterns are directly coded by genes; however, behaviour builds on what was selected for in the evolutionary history of the species.

Moreover, consider what Derek Bickerton has to say about evolution when he asks us to consider a species X with behaviour Y. He says that being a member of X gives X the potential to perform Y, but that capacity does not mandate that behaviour Y will be performed (Bickerton 2014, p. 52). The only reason that Y would be performed is if the members of the species who consistently perform Y contribute, at least, equally to the gene pool as those members within X who do not perform behaviour Y. Ultimately, it is the environmental pressures that will determine whether it is desirable or necessary to perform Y. While behaviour is certainly guided by the evolutionary process, this does not imply that we are under deterministic genetic control. Nevertheless, if we want to understand how humans were able to expand their capacity for 'thick cooperation', it *must* be based on an appreciation of human nature.

Social animals

All primates are deeply social, bonding together to form groups and remaining together over time. One important reason for primate group living is the benefit it provides for defence against predators. Being with others offers tremendous advantages in the search for food and the avoidance of being consumed as food (Wrangham 1980). The underlying argument is that 'a primate group is an implicit social contract (it is a collective solution to the problem of predation), and social contracts are always susceptible to being broken by free-riders, those who take the benefit of the contract but don't pay the costs' (Dunbar 2016, p. 38). In order for monkeys and apes to neutralize the stress that comes with group living, they form coalitions to take care of the free-riders and bullies. Dunbar argues that two processes are involved to 'create these bonds' (Dunbar 2017, p. 39). One is the unique role of endorphins in the maintenance of close relationships and the other is cognition, particularly mentalizing. I will have more to say about both these roles later in this chapter as well as the next two, but for now I want to suggest that primates as well as all mammals come together because of the neurochemical pathways that drive them to seek companionship. The suggestion that the primate group is a 'collective solution' to predation may be mistakenly interpreted as if it was some type of plan. I argue that while nonhuman mammals are predisposed to engage with others, they are not cognitively aware that they are doing so. It is not the case that the 'collective solution' came first and then the bonding mechanism of endorphins held the group together; rather, it was the neurochemistry that motivated animals to seek others that preceded the formation of the group.

The experience machine

In *Anarchy, State and Utopia*, Robert Nozick offers a helpful thought experiment by inviting his readers to imagine a machine that gives the possibility of experiencing anything that one would wish. This machine has the capacity to trick a person into having such experiences as writing a great novel or any other occupation that one's heart desires. The question Nozick asks the reader to consider is whether one would plug into the machine for life. After all, he challenges, 'what else can matter to us, other than how our lives feel from the inside?' (Nozick 1974, p. 43). If there is hesitation in plugging in and we choose not to, then Nozick suggests there must be something else that matters to us than just simply our experiences. He suggests that we want to do certain things and not just have an experience of doing them. Second, we also want to be a certain type of person, but Nozick counters that we can conceive of a transformation machine capable of transforming us into becoming anything we wish. Still, Nozick maintains, one would likely not plug in. According to Nozick, there must be something that does matter in addition to one's experiences. He concludes that perhaps 'what we desire is to live (an active verb) ourselves, in contact with reality' (Nozick 1974, p. 43). I argue this contact with reality referred to by Nozick is an engagement with others. We are programmed, so to speak, to engage with others.

Self-preservation

At the most basic level, organisms are in the business of self-preservation. Recall that in Chapter 1, I argued that the gene was the fundamental unit of evolution even though when speaking about evolution we generally tend to speak in terms of the individual. By using Dawkins' terminology, we found that in order for genes to survive over generations, it was necessary for them to construct successful vehicles to carry them. In order for those vehicles to be successful, they needed to be capable of 'conducting efficient life regulation' (Damasio 2010, p. 47). That is, successful organisms had genes that emerged by natural selection to provide them with a homeostatic network. This network includes such things as the self-regulating systems that are involved in the regulation of body temperature and blood sugar in mammals. For Damasio, in addition to the homeostatic structure, 'gene instructions led to the construction of devices capable of executing what, in complex organisms like us, came to flourish as emotions' (Damasio 2010, p. 47). Our emotional feelings are the result of the regions of the brain that are involved in our ever-changing internal states. An important feature here is that feelings are not only sensory but they also have an affective motivational element.

Seeking

Jaak Panksepp, a leader in affective neuroscience, maintains that there are seven main affective systems in the subcortical regions of all mammalian brains. He identifies these structures as SEEKING (expectancy), FEAR (anxiety), RAGE (anger), LUST (sexual excitement), CARE (nurturance), PAIN/GRIEF (sadness) and PLAY (social joy) (Panksepp and Biven 2012). He uses capitalized versions of these words to illustrate the distinct, ancient evolutionary neurochemical pathways and processes underlying the emotional states that we observe in animals and people. There is an impressive amount of evidence to suggest that 'primal emotional systems are made up of neuroanatomies and neurochemistries that are remarkably similar across all mammalian species' (Panksepp and Biven 2012, p. 4). These affective capacities provided animals with a selective advantage, as they efficiently anticipated survival needs. From this perspective, 'affects are in-built anticipatory neuropsychological mechanism of the brain' (Panksepp and Biven 2012, p. 43). These systems are comparable to homeostasis, the brain-based process that keeps our bodies in a steady state, adapting to the external environmental changes. When a homeostatic imbalance occurs, the organism feels unpleasant sensations, and similarly, organisms are negatively aroused with negative affects.

For Panksepp, the SEEKING system is at the root of all the other systems because it entails the positive feeling of anticipatory eagerness. He argues, quite reasonably, that SEEKING or expectancy is a much more accurate label than the traditional brain reward system because the seeking urge is fundamentally different from the pleasure of consummation (Panksepp and Biven 2012, p. 96). It is a seeking engagement with the world in order to find the resources necessary to prosper, as well as evade dangers or threats. It is not simply a reward system, and we saw the difference pointed out by Nozick between the experience in a machine and the active engagement with reality in seeking that experience. The SEEKING system is engaged automatically when the unpleasant states of homeostatic imbalances occur. Whether this system is called the reward centre or the SEEKING system, behavioural neuroscientists agree that it is fuelled by dopamine and that it is instrumental for learning. Whenever there is an increased release of dopamine in the brain, animals become more aroused. Panksepp suggests that the dopamine-energized SEEKING system is not simply a learning system; rather, it is a system that causes people and animals to engage the world. He argues that the initial predisposition, such as what we find with babies, is an 'unconditioned emotional affective response that is fundamentally independent of individual learning' (Panksepp and Biven, p. 136). I will provide evidence in Chapter 5 to suggest that babies are wired to engage with others.

Seeking mates

From the gene-centred point of view, the urge for sex does not play a crucial role in the survival of the individual but more importantly it increases the chances of passing on genes into future generations. Even though sexual gratification may not be essential for individual survival, there is evidence to suggest that people in well-bonded relationships are happier and feel more secure (Cacioppo and Patrick 2008). According to Donald Pfaff, all animals have an 'indubitable, irresistible' need to unite with their own kind (Pfaff 2015, p. 112). He correctly points out that this is the most elementary and primitive form of prosocial behaviour that fuels togetherness (Pfaff 2015). The building blocks for our prosociality are found in these neural mechanisms and the hormonal activity within our brains.

Vasotocin is an ancient hormone that ignites the sexual drive and birthing regulation in fish and reptiles. For instance, consider the increase in vasotocin levels when a female turtle makes the journey to a beach to begin digging a nest in preparation for the delivery of her eggs. Vasotocin reaches its peak when the last egg is deposited. As she covers over the last egg, her vasotocin levels begin to decrease and she then makes her way back to the sea, having no other interaction with her young. It appears that vasotocin and other related ancient neuropeptides evolved into 'oxytocin, which facilitates, not only female LUST, but is also a key brain system that promotes maternal CARE' (Panksepp and Biven 2012, p. 288). Now this may be a controversial suggestion, but the evidence supports the claim that the roots of caring feelings in all mammals come from these ancient circuits.

Research with small-brained mammals, for instance, suggests that both pair bonding and infant bonding are dependent on olfactory recognition information and tactile priming to release either OXT in females or AVP in males (Curley and Keverne 2005). The evidence suggests that the neural wiring that got animals together to have sex is at the root of our drive to seek mates and promote attachments. It will be argued that sexual and parental instincts are the foundation from which prosocial behaviours emerged. Bonding with a significant other and taking care of children are rooted in our genes, and this provided the capacity to care for others. In contrast, the description of the turtle laying her eggs provides a picture of a creature that does not care for her offspring. Mammals, on the other hand, must care for their offspring as well as feed and nurture them. This trait must have been selected for and this, I argue, provides the structure for our prosocial behaviour.

Kinship premium

Collaborating with Clive Gamble and John Gowlett, Robin Dunbar made the assumption that kinship played a significant role in hunting and gathering days, but presumed that today, we no longer rely on this type of favouritism and that we have come to value friends and acquaintances just as much as family (Gamble et al. 2014, p. 23). Yet, Dunbar's research points in a different direction, as our families still matter a great deal and, in fact, our extended family makes up about half of our social network (Dunbar 1996, p. 23). He provides additional evidence to show that those coming from larger families will list fewer friends on their web of relationships (Gamble et al. 2014, p. 23). In other studies, the researchers found that the closer people were to each other genetically, the more likely they were to provide a favour. This research led Dunbar to acknowledge that there is a marked difference between how cooperative we are with others compared with the special status we hold for kin.

Research with animals also points to the conclusion that kinship holds a special status for nonhuman animals as well. In studies with three mammal species and fifteen cooperatively breeding bird species, researchers found that helping behaviour was biased towards kin (Griffin and West 2003). Evidence from the last thirty years shows quite clearly that cooperation within animal populations is found typically with kin and seldom does it develop in groups of animals that are not related (Clutton-Brock 2009). Experiments with chimpanzees suggest that when chimpanzees assist others, it is almost exclusively with kin (Silk et al. 2005). Finally, although there is some evidence to suggest that primates will take care of abandoned neonates, these cases are very rare. In the natural environment, we can almost certainly conclude 'the death of a mother usually leads to the death of her infant or juvenile offspring' (Konner 2010, p. 434).

Researchers had similar findings while working with a colony of rhesus macaques just off the east coast of Puerto Rico, at Cayo Santiago. Investigators found that these primates formed into social groups, with a few adult females forming the nucleus of the faction. Rhesus macaques are philopatric, meaning the females stay in their natal group, while the males disperse (Berman 2016). Since the macaques were free from predators in this particular area, the researchers had the opportunity to monitor the behaviour of these primates who lived in groups consisting of four or five generations (Berman 2016). Over a three-year period, researchers found that female macaques showed a strong desire for kin and preferred close kin over more distant kin (Berman 2016, p. 67) The researchers suggested that 'the monkeys themselves recognized gradations of kin relatedness' (Berman 2016). They concluded that the behaviour of the macaques at Cayo Santiago supports Hamilton's rule.

There are a number of findings supporting the claim that primates come to sense a distinction between kin and non-kin. For instance, after two individuals are involved in a fight, and one of the two antagonists returns to the group and shows aggression to a third individual, it is likely that the victim will be a uterine relative of the original opponent (Cheney and Seyfarth 1986). While there is some evidence to suggest that primates make up after a fight (de Waal 2013), monkeys are more likely to make peace with uterine kin than with any other individual (Cheney and Seyfarth 1989). In other studies, it was found that nonhuman primates are disproportionately more tolerant of kin when they are near defendable food sources (Belisle and Chapais 2001).

It appears then that all primates, including humans, have what Dunbar calls a cognitive constraint on the size of social groups. Within that number, we have what has been termed a 'kinship premium' (Curry 2013b). That is, we tend to care more for those closely related to us. For instance, when it comes to donating a kidney, kinship is worth approximately 40 per cent more than one's relationship with mates (Curry 2013b). In another study, where participants had to endure pain to earn money for a particular person, the researchers found that participants were willing to sacrifice more for those who were very close to them (Dunbar 2013). This is significant because the person enduring the pain does so at a definite cost to the individual. The researchers suggest that even if the relationship with family members has been separated by time and distance, there is still a feeling of emotional closeness to a temporarily disconnected aunt or uncle (Gamble et al. 2014). Clearly, we have a kinship premium and we care more for relatives and provide assistance to them more than we do for others. The question to consider now is what are the psychological mechanisms dedicated to these positive feelings for helping behaviour?

Bonding mechanisms

Both oxytocin (OXT) and arginine vasopressin (AVP) belong to the neuropeptide hormone family that has a long evolutionary history of governing bonding and reproductive behaviours (Curley and Keverne 2005). For instance, vasotocin stimulates the nurturant moods and sexual behaviour in many species of birds (Godson and Bass 2001). It appears that vasotocin evolved into OXT and AVP, which are major players in the positive feeling that comes with nurturing a child. Dopamine, endogenous opioids, prolactin and other hormones are almost certainly involved as well (Panksepp and Biven 2012). All these mechanisms were not necessarily selected initially to serve social purposes, but their function seems to be the root of the mother's motivation to 'suckle, defend, and more generally, to devote herself to the welfare of her helpless juveniles until they were independent' (Churchland 2011, p. 32). The protective action by all mammals, including humans, seems to be regulated by emotion. The offspring who were cared for were more likely to survive, giving them a competitive edge for the continuation of the species.

Although different parts of the mammalian brain are involved, both OXT and AVP are key molecules in sex and social bonding (Insel 2010). Studies suggest that both molecules are important for pair bonding and mother–offspring bonding in a number of different species (McCall and Singer 2012). In research with the well-studied prairie and montane voles, evidence suggests that the former, a monogamous species, has a much higher density of AVP and OXT receptors in specific areas of the brain than the hermit-like montane vole (Lim and Young 2004). There is also evidence to suggest that other promiscuous species, such as Rhesus monkeys and the white-footed mouse, have similar types of OXT and AVP receptors as montane vole (Churchland 2011, p. 52). Behavioural biologists James Curley and Eric Keverne say that in addition to the more abundant distribution of receptors for AVP in monogamous species, just as importantly these receptors are more generous in parts of the brain that generate reward, 'thus facilitating the association between odour and the motivation for bonding' (Curley and Keverne 2005, p. 563).

The ability to bond with mate or offspring, or to prefer one partner over another, rests with the ability of the animal to form memories of their offspring or partner. There is evidence to suggest that OXT is linked to social memory in a number of species. In small-brained mammals for instance, OXT receptors are found in the parts of the brain that are linked to the process of olfactory information and in those parts generating reward (Curley and Keverne 2005, p. 563). In experiments with mice affected with mutated genes that encode for

OXT, researchers found that these mammals were unable to form olfactory memories of others (Curley and Keverne 2005, p. 563). This suggests that OXT is crucial for establishing infant recognition, as well as providing reward for identity and bonding.

The white-footed mouse and the montane mole are excellent examples of species that rely extensively on olfactory information in recognizing mates and infants (Zhang and Firestein 2002). In these species, a large number of genes encode for olfactory molecules (Zhang and Firestein 2002). In contrast, primates rely far less on olfactory information and distinguish between others by sight. Studies suggest that 'over 30 per cent of olfactory receptor genes are non-functional pseudogenes in non-human primates' (Curley and Keverne 2005, p. 563). In the human genome there are over 60 per cent of these nonfunctional pseudogenes (Curley and Keverne 2005, p. 563). Curley and Keverne hypothesize that the adaptive pressure to rely predominantly on sight rather than smell, due to the ecological change from a nocturnal to a diurnal style of living, led to the decline in olfactory processing. This led to the evolution of trichromacy in primates, which gave them the ability to interpret social information differently. For instance, 'colourful sexual adornments that signal reproductive receptivity in females and dominance in males led to more sexual interactions to be non-reproductive but provided for sexual bonding' (Curley and Keverne 2005, p. 563). The authors importantly add that mother–infant bonding extended beyond the suckling phase when facial recognition was necessary to keep track of mobile infants (Curley and Keverne 2005).

Curley and Keverne go on to suggest that the move away from a dependency on hormonal priming to the increasing use of higher cognitive capacities in social bonding of primates necessitated a reorganization of the brain. One of the consequences of this restructuring was that primates came to rely on facial expression to communicate their emotions. The authors argue that the executive brain has increased over mammalian phylogenies to a greater extent than the areas of the brain involved in hormonally regulated behaviour. For Curley and Keverne, the 'development of larger brains has, to some extent, released decision making from hormonal determinants' (Curley and Keverne 2005, p. 565).

Energy budgets and brain expansion

The question now is whether we can suggest the evolutionary pressures that gave rise to the runaway expansion in hominin brains, as it will be argued shortly that our greater cognitive capacity gave us the capacity to extend prosocial behaviour beyond kin to include others. There are two important points to keep in mind when considering the evolving brain. First, evolutionary change is guided by the necessity of the species to adapt to a changing environment. It is quite clear that environmental changes drive evolution. Some species, such as sharks, have remained stable for hundreds of millions of years due to an environment that has remained relatively unchanged. In contrast, land animals are frequently exposed to major, and at times quite dramatic, climatic changes; therefore, they must move locations, adapt to the environment or face death. The second significant point to keep in mind is that brain growth is very costly, so the advantages gained from the increase in size must outweigh the costs of making the changes.

Rather than suggesting that a higher intellect would have been favoured as those with bigger brains would have been more successful at hunting and obtaining food, Leslie Aiello and Peter Wheeler stressed the importance of the quality of the diet. They reasonably argued that brains are greedy for energy, and they found across the primate line that some had smaller guts than others. From their research, we can conclude that relative to body size, those primates who had larger brains had the smallest guts. For Aiello and Wheeler, the Expensive Tissue Hypothesis states that large brains are made possible when less energy is spent feeding intestine tissue and more is spent on brain tissue (Aiello and Wheeler 1995).

As mentioned, the development of trichromatic colour vision in the ancestor to Old World monkeys and apes led to these species being able to recognize fruit and thereby consume fruit rather than leaves. The colour of fruit illustrates the ripeness, as well as its nutritional value. Aiello and Wheeler argue that it takes much more energy to digest leaves as opposed to ripe fruit, and the difference in brain size is correlated with the difference in energy budgets. They point to evidence suggesting that primates relying mainly on fruit have larger brains than those species relying mainly on leaves (Allman 2000). Biologist John Allman hypothesizes that an animal may use more of its energy to eat the abundantly available, less-nutritious leaves or the animal may use the 'available watts to support an enlarged brain with the capacity to have information about the location and cognitive strategies necessary to harvest food that is scarce and hard to find' (Allman 2000, p. 167). Nevertheless, a problem arises when the targeted food becomes harder to find.

As long as there is plenty of fruit available for both apes and Old World monkeys, there is no pressure for either species to change their behaviour. Since great apes do not have the ability to digest unripe fruit, they would eventually perish if all the fruit was eaten before it had time to ripen (Allman 2000). Old World monkeys, on the other hand, do have the necessary enzymes capable of breaking down the tannins in fruit, so they have the ability to eat unripened fruit. Consequently, about 10 million years ago, as Africa became drier, the Old World monkeys had an advantage over apes and would have become the dominant primate, forcing the apes to the fringes of the forest. However, one lineage of apes, the one that led to the species *Homo*, managed to survive, but like apes, *Homo* did not have the accommodating digestive system. This forced them to find different food sources.

Hunting hypothesis

There is little agreement among anthropologists as to whether early hominins acquired food from hunting or scavenging. Indeed, an influential and long-established model in anthropology has been the hunting hypothesis. However, James O'Connell and colleagues offer reasons to be sceptical of that hypothesis (O'Connell et al. 2002). The hunting hypothesis suggests that with the expansion of the open savanna in Africa, there would have been less fruit available for early hominins, but that there would have been an increase in large herbivores as a potential food source (O'Connell et al. 2002). This increase in available meat is due to the fact that wherever grass grows, large animals thrive. Since grass grows from the root, it generally does not get killed from grazing, and this results in the availability of plenty of food for the animals. The hunting hypothesis suggests that in order for males to provide for the family, they would have had to cooperate together to be successful in hunting large herbivores. This, in turn, would have led to pressures for larger brains essential for increased cooperation and those successful hunters would have passed on their genes to their children.

A major problem with the hunting hypothesis is the lack of archaeological evidence to support the claim that hominins were armed with the necessary means to take down large predators. The projectile-type weapons or bows and arrows required for a successful hunt appear much later in the archaeological record (O'Connell et al. 2002). The weapons that have actually been found suggest that they were used for scavenging, rather than hunting. This becomes clearer if we consider the distinction between passive and aggressive scavenging. The former, as the name implies, is the 'culling of scraps from carcasses heavily ravaged and abandoned by their initial non-hominin predators' (O'Connell et al. 2002, p. 832). Aggressive scavenging, on the other hand, is the taking of prey from the original predator(s) before the original predator(s) would have normally abandoned it. This suggests that acquiring edible tissue is highly variable, but one can assume that at times much of the carcass would have been available with much of it intact (O'Connell et al. 2002).

O'Connell and colleagues admit that there were some methodological problems accounting for the damage pattern counts at a number of archaeological sites. However, they do suggest 'aggressive scavenging by hominins on carcasses variably ravaged by carnivores might well generate the damage patterns reported' (O'Connell et al. 2002, p. 848). Since these hominins lacked the necessary weapons to be hunters, they would have had to cooperate in large numbers in order to frighten competitors. If they were unable to intimidate predators, they would not have been able to have access to the meat. Attracting others to help fight off predators would have required recruitment, and recruitment calls for cooperation, and since we know that primates don't generally cooperate in this fashion, there must be an alternative explanation for recruitment. There are a number of plausible hypothesis put forth to account for the need for this type of cooperation, and they all deal with the emergence of language.

Now there are a number of different theories to account for the evolution of language. For instance, some have argued that language evolved as a costly signally device to enable females to assess male quality (Miller 2001). Others suggest it was to promote pair bonding (Deacon 1997), or to facilitate toolmaking (Greenfield 1991). Yet others suggest that language evolved to assist mother–infant communication (Falk 2004) or as a substitute for grooming (Dunbar 1996). Furthermore, there is no consensus as to when language first evolved. Incredibly, there is a 2-million-year gap between estimates of language emergence by the experts. Terrence Deacon and Derek Bickerton place language emergence around 2 million years ago, while others suggest it took place closer to 50,000 years ago. Robin Dunbar places language emergence closer to the middle of these two extremes. I suggest that the ecological niche in which hominins found themselves in at the time of aggressive scavenging put selective pressure on some form of oral communication. While I find Bickerton's account of language evolution the most compelling, I will begin with Robin Dunbar's influential hypothesis.

Dunbar reasonably points out that predation is a significant evolutionary problem. Those species that solved this dilemma were able to breed and reproduce, passing their genes on to future generations, thus solidifying the success of the species. Predation is not as much of a problem for those large and powerful species, such as chimpanzees and gorillas, but for early hominins, predation was a major risk (Dunbar 2016). Dunbar suggests that one way of reducing the risk of predation is to live in large groups. Indeed, one of the distinguishing features of all primates is that they do indeed live in groups. He points out that 'primate groups have a continuity through time, a history built on kinship' (Dunbar 1996, p. 18). This is one of the main arguments in this chapter – that kinship is indeed the foundation of human cooperation – but as we will see, humans were successful at expanding cooperation outside of the kin group to include strangers. Nevertheless, the means by which primates used to hold their coalitions together, according to Dunbar, resides in brain mechanisms that are stimulated by touch.

Grooming

Old World monkeys and apes spent a great deal of time grooming each other (Dunbar 1996). Grooming stimulates the production of endogenous opiates. The feeling that is caused by repetitive stimulation is comparable with opiate drugs, such as opium or its derivative heroin, and the 'high' experienced by long-distance runners (Dunbar 1996). Dunbar suggests the good feeling derived from the grooming may be the mechanism that encourages the grooming, but this may not necessarily be the evolutionary reason as to why they do it. He suggests that the selection pressure had more to do with cementing bonds between grooming partners. There is some evidence to suggest that alliances are maintained by grooming. For instance, in studies with vervet monkeys, Robert Seyfarth and Dorothy Cheney recorded the screams by vervets that were attacked by other members of the group. The researchers then played the recorded screams back to see what the reaction was of other monkeys. They found vervets were only interested in the screams from those who had groomed them during the previous two hours. This suggests that the monkeys were differentiating between grooming and non-grooming partners and the grooming partner was someone special (Dunbar 1996). For Dunbar, grooming is the glue required to hold coalitions together.

Machiavellian hypothesis

According to Niccolo Machiavelli, it would be in the best interest of the prince to get the support of the common people and not necessarily the nobility because the latter are so close to the position of the prince that they could essentially taste the power. The cunning author also advocated life strategies that were devious in order to obtain and keep power. Apparently this advice is good for all primates as among male chimpanzees, power is something to strive for, and when they form coalitions, they do so because they must in order to survive. Two British psychologists, Dick Byrne and Andrew Whiten, found monkeys and apes were especially good at using social knowledge about the other monkeys in the group and used this information to seek power. Like a true Machiavellian, the primates mischievously used information to deceive and outwit each other. For instance, monkeys were found to hide desirable food from others and participate in a sort of troglodyte distraction theft by making alarm calls to distract the rest of the group in order to capture a meal for themselves (Gamble et al. 2014, p. 17). The psychologists aptly called this the Machiavellian Intelligence Hypothesis.

Social brain hypothesis

Influenced by Byrne and Whiten, Dunbar realized that there was a link between the size of brains and the size of coalitionary or social units. Dunbar's Social Brain Hypothesis suggests that the primate brain evolved in order for the animals to participate in a complex social world, and Dunbar has shown that the size of a species' social group correlated with the size of the neocortex. The neocortex, the part of the brain responsible for complex thought that includes such things as hypothetical thought and scientific reasoning, has expanded the most in primate evolution, and it is due to this part of the brain that primates have the largest brains of all mammals (Allman 2000).

Dunbar plotted the group size of various species of monkeys and apes with the relative neocortex size of the species. The equation that he derives from this graph predicts that the natural group size for humans is 150. This has come to be known as Dunbar's number. He provides considerable evidence to support this conclusion by showing that typical hunting and gathering societies are organized as a hierarchy, that is, 'families clustered within kinship groups, kinship groups clustered around villages and villages clustered within larger regional groups' (Gamble et al. 2014, p. 21). It is the last level of social organization that is interesting for Dunbar and colleagues, as it hovers around 150, just as his hypothesis predicts. Dunbar's number and parts of the Social Brain Hypothesis make sense; however, problems arise when he applies this to account for the evolution of language.

The Social Brain Hypothesis suggests that the tremendous increase in brain size correlates with the change in group size. From this, Dunbar and colleagues argue that the size of the group is what determines the need for language. These researchers are suggesting that as groups got larger, there came a point in time where grooming had to give way to language, as this form of cohesion would not be able to accommodate the increase in social partners. The implication is that at some point in our evolutionary history, it was necessary for grooming to give way to language.

What are you saying?

The question that comes to mind when contemplating Dunbar's account of language is, what did these early hominins talk about? According to Dunbar and his colleagues, early hominins talked about each other. For them, humans are clearly in the business of gossiping, and they see no reason why it would be any different half a million years ago or earlier than from today (Gamble et al. 2014, p. 143). They suggest that 'language is our ancestral way of learning about others and influencing them to sign up to our social projects' (Gamble et al. 2014, p. 143). They go on to speculate that the conversation may be as simple as deciding what to eat and with whom to dine (Gamble et al. 2014, p. 143). The problem with this conclusion is, as they themselves suggest, '(just) as brain size changed *gradually*, it is very likely that language evolved gradually' (Gamble et al. 2014, p.142, *my italics*). If language evolved gradually, as one could only imagine, one would presume that the first words were grunt-like so it would be impossible for early hominins to be able to have conversations or influence others to sign up for social projects. Doesn't convincing others of one's position require a fully functioning or at least a semi-functioning language? One has to be able to already have a full language if one wants to gossip about others. How does hominin A warn hominin B that hominin C is a lazy or selfish brute if language is in its infancy? There is evidence to suggest that hominins at this time were anatomically capable of making sounds. This is indicated by brain configuration, 'particularly a section associated with speech called Broca's area, and a likely position of the larynx low enough to permit some degree of articulate speech' (Sale 2006). There is no question that language provides for everything that Dunbar and colleagues suggest, but the evolutionary pressure for language to evolve must have come from something other than a substitution for grooming. In order for all living creatures to survive, they need fuel, and since animals, if hungry enough, will even give up sex in order to eat, then perhaps the selection pressure for vocal communication came from the circumstances in the search for food. Derek Bickerton offers, arguably, the most plausible scenario to date for the evolution of language and more importantly for this chapter, his hypothesis provides evidence to show how it could be possible for humans armed with the tools of a fully functioning language and symbolic thought to evolve the capacity for greater cooperation beyond kin to include strangers.

Putting-the-baby-down hypothesis

As I have argued, it is unlikely that it was either hunting or gossiping that selected for language. While I have argued in this chapter that the mother–child relationship is definitely special, it is unlikely that language evolved from this relationship either. Dean Falk offers such an explanation, and while her hypothesis seems plausible, it doesn't clearly lay out the reasons why language is necessary for the circumstances she describes. In her putting-the-baby-down hypothesis, Falk contends that since human babies are helpless, they need to be cared for by the mother for a greater amount of time than any other primate (Falk 2004). Falk reasonably suggests that in hunting and gathering peoples, while the mother was gathering, she had to put the baby down because it would be impossible to gather food while holding the baby. Babies do not like being left alone, and so to reassure the baby that the mother is still close by, she would have to resort to some type of verbal communication. According to Falk, words would have inevitably emerged and language would have evolved. There seems little doubt that the caring for children played a role in cooperation, and I will speak to that shortly, but I am less convinced that the relationship was the stimulus for language. There must have been something else going on.

Bickerton counters Falk by suggesting that if we want to know how words 'emerge', we need to identify the circumstances as to why there would be such a need for words. It would seem that perhaps humming or some other soothing sound may have been enough to keep the baby satisfied for the particular situation described by Falk. This type of communication is similar to the type of communication systems used by animals, and since no other animal has evolved language, then there must have been different pressures to select for language. Therefore, as Bickerton suggests, there is a need to discriminate between Animal Communication Systems (ACS) and language.

More than animal communication systems

Bickerton maintains that communication used by animals is only concerned with survival, mating and social signals. Warning or food calls, for instance, are beneficial from an inclusive fitness point of view, as helping family members will give a greater chance of passing on genes. After all, evidence suggests that 'if they (nonhuman primates) are not with close relatives, they are less likely to call than if they are with immediate family' (Bickerton 2009, p. 19). Second, all ACS are grounded in the here and now. Animals can't, as far as we know, caution conspecifics by reminding them about the predator who was hanging out at the local watering hole, last weekend. Humans can communicate about last week, next week, things not here and even about things we have never seen before. Szabolcs Szamado and Eors Szathmary say the claim that language originally evolved to replace grooming or facilitate mother–infant bonding does not meet the criterion of conveying information that is not connected with the observable present (Szamado and Szathmary 2006).

Renowned primatologist Richard Wrangham says the reason horses eat grass is not because they have the right kind of teeth and guts for consuming grass; rather, they have the type of teeth and guts they have because they were adapted to eating grass. He goes on to say that humans don't eat cooked food because we have the type of teeth and guts to accommodate cooked food; rather, we have the type of teeth and guts because we adapted to a cooked diet (Wrangham 2009, p. 89) Similarly, humans don't have language because we have the type of big brains that could accommodate language but we have the type of brains that we do because we adapted to use language. What then could have been the selective pressure for humans to evolve language that goes beyond a particular situation and the here and now? After all, ever since there was life on this planet, no other animal has needed to deviate from the standard communication system. Bickerton says that the essential property of language is symbolic, rather than indexical, and this requires the property of displacement. According to Bickerton, displacement is defined as 'the ability to talk about things that are not physically present' (Bickerton 2014, p. 82). This is a crucial point for any account of language.

For Bickerton, language arose not through unmotivated genetic changes but as a direct result of a specific 'ecological problem that – one ancestral to humans 2 million years ago – had to face' (Bickerton 2014, p. 11). This ecological niche that early hominins found themselves in must have put pressure to transmit information about things in the environment that were not immediately present.

As mentioned earlier, hominins were forced out from the forests to search for food. Furthermore, the situation had to be such that the information to be passed from one hominin to another was informative. After all, human language is essentially informative and this is manifestly different from what is found in ACS, as animal communication is essentially manipulative (Krebs and Dawkins 1984). Of course, humans have the capacity to be manipulative as well; however, language is essentially informative.

Rather than comparing human language with the communication systems of our simian cousins, Bickerton cleverly invites us to consider other species for which there are signs of mechanisms of displacement, such as what is found with the hymenoptera. Granted there is quite a phyletic distance between humans and ants and bees, but this should not deter us from considering homologies across these species. For instance, we see the sameness in the bat's flying wing and the grasping hand of a human. A second objection to using hymenoptera for comparison with humans is the fact that bees and ants operate instinctively; however, Bickerton rightly points out that language is a 'brain internal physical mechanism that lies far beneath the reach of conscious cognition' and is just as instinctual as what the bees and ants do (Bickerton 2014, p. 82). What is most important for our purposes when considering communication systems is to identify the problems faced by ants and bees that led them to evolve mechanisms of displacement and, then from there, compare that with similar problems faced by early hominins but not faced by other primates.

What were the food sources, for both ants and bees, that would require the need to inform conspecifics about a potential food source? Ants are attracted to dead organisms, while bees obtain their food from flowering plants. For both ants and bees, time is of essence. Their source of food is only available for a certain amount of time, so they have to recruit others and collaborate in order to take advantage of the time-sensitive food source. With bees, in order to inform others, they instinctively return to the hive and perform a 'dance' for nest mates. If the food is within a distance of say 100 yards, then the bee does what is termed a round dance, and if the food is further afield, then the bee performs elliptical loops, known as a waggle dance (Bickerton 2009, p. 133). Ants, on the other hand, while foraging for transient and scattered food sources, communicate chemically rather than engage in a dance performance. After locating a food source, the ant returns to the nest and leaves a chemical trail to 'show' where to find the food. Those bees and ants that practised this form of food sharing would have a greater prospect of survival, and genes that supported these means of communication would spread throughout the population.

Recruitment

The recruitment of nest mates by bees and ants is comparable to the problems our hominin ancestors faced when an individual came across large fauna. There would have been a need to recruit others to help with fighting or scaring away predators. All the evidence points to early hominins having a small stature (Johanson and Edgar 1996). They were also under pressure to act quickly. Hominins had to recruit by telling others that there was a type of object in a specific place. They had to recruit and work together in order to obtain food or they would perish. These ecological conditions would have put pressures on early hominins to somehow identify the object. As far as we know, it is only humans who are capable of naming things, and the argument here is that Bickerton offers us the most plausible scenario as to how and why early hominins needed to identify and remember an object, then recruit and tell others about that object.

There is a fascinating YouTube video about Christian the lion who, as a pup, was raised by two men. Understandably there comes a time when the lion becomes just too big for human interaction and eventually Christian was transported to Africa and set free. About one year later, the two owners went to Africa to see Christian, even though they were warned that Christian was now even larger and wild. Once Christian was reunited with his human caregivers, it appears that the lion had remembered them. Now this leads to the question as to whether, prior to this reunion, Christian actually thought about his human caregivers. Did the lion wonder, or have the capacity to wonder, about those two caregivers when they were separated? Unlikely, as the evidence we have about animal cognition suggests that they live in the 'remembered present' (Edelman 2004). As far as we know, animals are only concerned with things that are in their vicinity.

The ability to think of objects that are not immediately present would seem to be the stepping stone for early hominins to have the capacity to imagine possible outcomes. Clearly the displacement theory of human language provides us with a plausible description of the necessary steps that led early hominins from recruiting others to the need to use a symbol or words to identify a specific object that was not immediately present. Furthermore, recruiting others would necessitate skills for forming a joint goal to work together with a partner in their personal common ground (Tomasello 2014). For Tomasello, this common ground led each participant to recognize that each individual has a role to play and that each was of equal value (Tomasello 2014). Those individuals working together on a joint commitment were responsible to each other and each partner was dependent on the other for survival. This would lead individuals to have a shared intentionality (Tomasello 2014; 2016). It would also put pressures on individuals to demonstrate their potential as good collaborators. I will have more to say about commitments in the next chapter, but for now I will comment on Tomasello's concept of shared intentionality.

Shared intentionality

Michael Tomasello's Shared Intentionality Hypothesis suggests that due to the problems associated with individuals attempting to collaborate and communicate with others, humans evolved thinking processes that included representation, inference and self-monitoring. He says that the likely first steps occurred around 2 million years ago, when early hominins created new forms of social coordination in the context of collaborative foraging (Tomasello 2014). He has come to the conclusion that the crucial steps in the evolution of human cooperation had to do with how humans obtained food. For Tomasello, when hominins engaged in an activity that provided a mutual benefit, either by physically helping the other or by informing them, each one is helping themselves, as their success of the role they play is essential for the overall success (Tomasello 2014). In order to provide evidence for this unique human ability, Tomasello takes a comparative approach. He compares the cognitive capacities of primates with humans, besides comparing the cognitive capacities of children across different ages. I will have more to say about the experiments with children in Chapter 5, but for now, I will concentrate on some of the work Tomasello conducted with primates.

Tomasello acknowledges that all social animals collaborate with each other as they typically forage together in a group. He wants to determine, however, what sets humans apart from other primates in collective activities, and he suggests that by comparing the different hunting styles of humans and chimpanzee, it will provide us with clues to determine the difference. According to Tomasello, when chimpanzees hunt red colobus monkeys in the wild, the chimpanzees appear to have a shared goal and seem to partake in a division of labour (Tomasello 2009). He says that typically during the hunt, one chimpanzee gives chase, while others climb trees in order to block the monkey from changing direction and others are tasked with circling around in order to block any chance of escape. The remaining chimpanzees position themselves on the ground just in case the monkey falls down. For Tomasello, although it may appear to be the case, there is no prior plan; rather, each chimpanzee is simply maximizing its best chance of catching the monkey.

Philosopher Raimo Tuomela calls this 'group behaviour in I-mode' (Tuomela 2002). Tuomela explains social behaviour by describing the 'we-mode' attitude as one that involves thinking and acting from the perspective of the group. All the members of the group are collectively committed to this mental outlook. The 'I-mode', on the other hand, lacks these two characteristics, and the action of the individual is strictly for a self-directed benefit. According to Tuomela, social practices can occur in either mode; however, the 'we-mode' practices are fundamental for institutional practices. I will have more to say about institutional practices later.

Another example of group behaviour in 'I-mode' is when a group of monkeys come across a fruit tree. They all scramble up the tree, grab their fruit and eat it, on their own. This would be a non-social collective action on the part of the monkeys. Tuomela calls attention to Max Weber's example of a number of individuals on the street, and once it rains, they all simultaneously open their umbrellas. With the example of the monkeys coming across a fruit tree and the people opening their umbrellas, we see that the individuals do not take into account the thoughts and actions of others in order to partake in group behaviour. I argue that only humans have this ability, and this is the difference between collaboration and 'thin cooperation'.

Tomasello says that when comparing human behaviour with nonhuman primate behaviour, it is clear that the evidence suggests that apes behave individually, while humans behave (mostly) cooperatively. For instance, he notes that humans, but not apes, engage in cooperative childcare. Second, human communication is informative, while ape communication is essentially coercive. Third, humans, but not apes, teach each other helpfully, and finally, it is only humans who establish social norms and institutions (Tomasello 2009; 2014; 2016). Tomasello concludes that humans are fundamentally cooperative, while apes are mainly exploitive.

Where Tomasello and Bickerton part company is in the former's insistence that language is 'the capstone of uniquely human cognition and thinking, not its foundation' (Tomasello 2014, p. 127). For Tomasello, language is not the source of unique human cooperation but is the result of it. Tomasello's 'shared intentionality' rests on the ability to create with others joint intentions and joint commitments (Tomasello 2009; 2014; 2016). In order to make joint commitments, it must be based on processes of joint attention and mutual knowledge, and this requires the capacity for representation and inference drawing. Both authors are essentially talking about the same time period and the same selective pressures but differ fundamentally about the role of language. Nevertheless, Tomasello makes it clear when he states, 'this novel form of moral psychology was not based on the strategic avoidance of punishment or repetitional attacks from "they" but, rather, on a genuine attempt to behave virtuously in accordance with our "we"' (Tomasello 2016, p. 5). Furthermore, it was not until hominins evolved the necessary cognitive skills for language that there was the capacity to create cultural norms and conventions. The ability to create norms and conventions elevated human cooperation to unprecedented levels, giving humans the capacity for 'thick cooperation'.

There are others who have argued that language evolved in humans as a trait for promoting greater cooperation (Pagel 2012), while some have added that language evolved as a specific form of cooperation, namely, teaching (Laland 2017). No question that teaching is important, but again other animals teach their offspring without language, so we need evidence of what environmental conditions would have led humans to select for language. The steps for the evolution of language, offered by Bickerton, seem to be the only one that accounts for displacement, and as I have mentioned, this is crucial for explaining how humans think, speak and consequently have the capacity for 'thick cooperation'.

Competence and comprehension

Perhaps Daniel Dennett's distinction between competence and comprehension can be useful here. In the opening paragraph of his latest book, Dennett says that we know that there are bacteria, but animals such as dogs, dolphins and chimpanzees do not. Bacteria don't even know that they are bacteria (Dennett 2017, p. 3). He goes on to suggest that nevertheless bacteria are competent to 'respond to other bacteria in bacteria-appropriate ways ... without needing to have any idea about what they are doing' (Dennett 2017, p. 80). For every living thing on this planet, except us, competence without comprehension is the way of life. When biologists give reasons as to why animals behave in certain ways, Dennett says that it is fine to adopt the intentional stance, but biologists cannot attribute any comprehension to the organism. For Dennett, animals, plants and microorganisms are competent to deal with what is going on in their environment and there are a host of *free-floating rationales* to account for these competencies (Dennett 2017, p. 80). By using this clever phrase, Dennett is suggesting that natural selection is an 'automatic reason finder' in that while natural selection does not have a mind to design, it is competent to perform 'this task of design refinement' (Dennett 2013, p. 234). To make sense of the idea of *free-floating rationales,* Dennett invites us to think of the terms *numbers* and *numerals*. The latter can be Arabic or Roman, but the point is that they are human inventions, while numbers are not. *Free-floating rationales* are like numbers and not numerals. The important point here is that comprehension resides with humans alone and the mind tools necessary for comprehension evolved once humans had language.

The major claim in this chapter is that at a particular period in our past, biological evolution got us to the stage where there was a capacity for protolanguage. As language evolved, the human unique culture developed, and once that happened, the influence of our genes on our behaviour was modified and we had the capacity to override some of those influences. Once there is comprehension in the way Dennett describes, humans had the ability to determine why they are being motivated in a certain way, that is, reflect on the motive and consider whether they ought to act in this way. Of course, this is a very general framework, and it is very difficult to give a specific timeframe. I am not suggesting that at Time A we didn't have language, and then at Time B, we did. Think of the first word spoken by an infant, when he or she says *dada* or *mama*. We don't generally say that the infant is speaking a language and it is difficult to pick a specific day when we can say today she knows language but yesterday she did not. The same can be said about language evolution – it was an extremely long drawn-out affair.

It seems to me that Dennett would be in agreement with this, but he spends a great deal of time defending memes and he rationalizes this perspective-taking approach because, unlike other cultural accounts, he claims the meme's perspective allows us to see how 'culture-borne information gets installed in brains without being understood' (Dennett 2017, p. 213). For Dennett, the problem with the standard view is a reliance on an uncritical assumption that comprehension is the source of competence and only with a meme's perspective can we 'account of how uncomprehending competences can gradually yield comprehension' (Dennett 2017, p. 213). The chief benefit of meme's-eye point of view, according to Dennett, is to give us the competence to ask questions, such as 'is x the result of intelligent design?' or 'is x a good worth preserving and bequeathing or a bit of parasitic junk?' (Dennett 2017, p. 247). I don't see any problem identifying 'tunes, ideas, catch-phrases, clothes, fashions, ways of making pots or of building arches' as memes (Dawkins 1989, p. 192). However, the question to be asked is, what additional explanatory power does one get from taking the perspective that it is the memes themselves that benefit? Many of our ideas throughout history, both good and bad, were unintentional, but why do we need a meme's point of view to admit this? Certainly many ideas that shaped civilization were the result of a bottom-up process without comprehension, but much 'progress' stems from ideas directed by human brains. The ideas of human rights and autonomy were 'directed' and 'endorsed' by human brains, and I argue it was these ideas that paved the way for 'thick cooperation'.

Creating facts

We have seen with some of the studies with chimpanzees that they have the ability to perceive, remember and desire things, and we can also imagine that early hominins would have initially had the competence but not the comprehension to do the same. Also, early hominins would have had the capacity to discriminate between different features of the environment. In our scenario of scavenging hominins, we would suspect that they would have been competent to differentiate between animals, but they would not have had the capacity to have concepts such as identifying a particular animal. Additionally, they would certainly have understood agency. John Searle argues that what is not possible with prelinguistic intentional states is the ability to 'create facts in the world by representing those facts as already existing. This remarkable feat requires a language.' (Searle 2010, p. 69).

When Dennett suggests that biologists can give reasons as to why things are in a certain way, he is saying humans can comprehend functions. For instance, in the seventeenth century, scientists discovered that the function of the heart is to pump blood. According to Searle, once we discover that the function of the heart is to pump blood, we assume that survival is a positive value and subsequently by describing something in terms of function we can introduce a normative vocabulary (Searle 2010, p. 59). Searle says we can say that a particular heart is suffering from disease, but we cannot say the same things about a stone suffering from a disease. However, once we assign the stone a function, such as this rock in my hand is a weapon, we can make evaluative claims about the rock as a weapon. For Searle, language is the prerequisite to having a moral concept.

Function of language

If we agree with Bickerton's claim that the initial selection pressures for language were to tell others about the location of food, we see that the primary function of language was to convey honest and accurate information. Searle argues that if a person intentionally conveys information to another person, then that speaker is committed to the truth of the fact. The speaker, in other words, has made a public performance (Searle 2010, p. 83). This act goes beyond 'the commitments of the intentional state expressed' (Searle 2010, p. 81). Searle reasonably argues that there is a huge difference between a privately held belief by a person and the expression of that belief to others. They both involve commitment to the truth of the claim; however, when publicly expressed, the speaker additionally has to provide reasons for the belief and can be held accountable. For Searle, once we have language, we have a deontology because there is no way of making speech acts without making a commitment. So what is different from what we have with prelinguistic intentionality is 'the public assumption of conventionally encoded commitments' (Searle 2010, p. 84).

Searle goes on to argue that once we have a deontology that is collectively created by the intentional action of individuals, it can be extended to social reality generally. He says that humans, because of language, can not only represent our preexisting states of affairs to others but we also have the additional capacity to 'create states of affairs with a new deontology; we have the capacity to create rights, duties and obligations by performing and getting other people to accept certain sorts of speech acts' (Searle 2010, p. 85). Language is not just about describing states of affairs or about allowing us to gossip about others or for reassuring our children; rather, language importantly gives us the capacity to create. We can create special status functions by declaring what Searle terms X-counts-as-Y-in-context-C. In order to wrap up this part of the argument, I will spend some time unpacking the main concepts used by Searle.

One of the main distinctions between humans and other animals is our capacity to appoint functions for objects and people. Searle calls these status functions. According to Searle, status functions are the glue that holds society together. Additionally, the ability to perform this function rests on the fact that it is collectively recognized that the object or person has the status to carry out the said function. An example would be a police officer. A police officer can carry out certain functions in virtue of the fact that it is collectively recognized that the police officer has the status to carry out these special functions. It should be made clear that acceptance does not necessitate approval. One may not approve of the fact that the officer handed her six traffic tickets but recognizes the fact that the officer has the status to do so. Second, status functions are dependent on collective intentionality. The only way that I can go to the mall, park my car between two parallel yellow lines and buy lunch with a couple of pieces of paper or plastic is by virtue of the collective recognition that the piece of paper or plastic represents money, that the car is my property and that I am parking in a space reserved for vehicles.

What John Searle so clearly states is that with language, humans, and apparently humans alone, have the ability to create a reality by simply declaring that the reality exists. So we can create money and many other things by simply representing them as such. Once we create these conventions, and although they are arbitrary, once they are in place, all members have a right to a number of expectations. It is these rights and corresponding responsibilities that provide humans with the capacity for 'thick cooperation'. I will have more to say about institutions in the next chapter, but for now, I will argue that the hypothesis suggesting that the pressures for communication associated with scavenging (that required such things as memory, planning and the intentions of others) put pressure on the hominin brain. While there is the demand for an increasing brain due to language acquisition, there must be an accompanying evolutionary explanation as to how hominins got the fuel to support such an expansion. It seems cooking offers a reasonable explanation to account for how hominins obtained the necessary fuel.

Psychologist Suzana Herculano-Houzel argues that the human advantage is not simply brain size, as elephants have larger brains than humans; rather, it is the fact that humans have the largest number of neurons in the cerebral cortex (Herculano-Houzel 2016). This, she suggests, was the result of the discovery of cooking approximately 1.5 million years ago. Richard Wrangham argued that the hominin brain increased so rapidly due to the larger caloric intake of cooked food. He points to evidence suggesting that since hominins no longer had to expend energy on hard-to-chew food, there was a reduction in tooth and cranial bone mass (Wrangham 2009). Second, Wrangham says that there is evidence of cooking as early as 1.5 million years ago. Although the date when hominins discovered cooking remains controversial, an international crew of scientists found evidence of hominins building and tending to fires deep inside the Wonderwerk Cave in South Africa (Berna et al. 2012). These scientists discovered bones and bone fragments that showed evidence of having been burned, suggesting that hominins were cooking over 1 million years ago (Berna et al. 2012). Finally, Herculano-Houzel provided evidence that suggests the typical raw food diet of primates is so limiting that without a change in diet 'the evolution of modern humans would simply not have been possible' (Herculano-Houzel 2016, p. 193).

If we consider that it took 50 million years for the primate brain to increase by about 29 billion neurons and within a relatively short period of time – 1.5

million years – the *Homo* brain added 57 billion more neurons, there must have been something allowing the brains of *Homo* alone to increase to such an extent (Herculano-Houzel 2016). To achieve this increase, hominins had to decrease either the size of the body or the energy cost of the brain or increase the amount of time foraging and hunting or increase the amount of energy from the same amount of food (Herculano-Houzel 2016, p. 189). The solution for hominins was the latter, as cooking increases the caloric yield of foodstuffs and decreases the energy and time required to chew the food to obtain those calories (Wrangham 2009). With additional time available, hominins had the ability to look for better ways of gathering and navigating the environment. Additionally, the demands of language and cooperating with others would have put pressure on the increase in neurons in the brain. Before wrapping up this chapter and concluding that kinship is the foundational organizing principle of early human societies, I will look at what Sarah Bluffy Hrdy calls cooperative breeding (Hrdy 2009).

Bonding

Some species bond quite quickly and exclusively with their offspring, while there is a larger window of opportunity for others. For instance, precocial species such as sheep bond quite rapidly with their young. These animals are born with the capacity to move quite freely, so there are selective pressures to bond quickly with the mother since there is the risk of getting lost. Humans, on the other hand, have a notable large window for social bonding. Like other altricial species, human babies are incompetent to care for themselves and are entirely dependent on caregivers. While human babies usually bond with their mother, they are unique in the fact that they readily bond with other caregivers as well. Therefore at a very young age, human babies are more receptive to bonding with a number of strangers.

Cooperative breeding

Notwithstanding the importance of the mother–offspring dyad in most species, Sarah Bluffer Hrdy (2009) questions whether the focus on this relationship has taken away from the importance of others in the raising of young. Maternal care, without doubt, is the costliest investment for the mother and therefore she could use help from others. Hrdy argues that in the evolution of hominins, alloparents such as fathers, siblings, aunts, uncles and grandmothers were instrumental in helping with care. The family enterprise of helping with the raising of their young leads Hrdy to characterize humans as cooperative breeders (Hrdy 2009). A cooperative breeding environment is a necessary habitat for the slower maturation of babies, and this would also allow human mothers to give birth to another baby before the other infant is weaned. Within all primates, only human mothers give birth to offspring within such close succession (Kaplan et al. 2000). So with more than one child to look after, the mother would need help from others.

If the species leading to *Homo sapiens* lost their hair over 3 million years ago, infants would have been incapable of hanging on to the mother, so it seems probable that the mother could have used an extra set of hands to hold the baby when she was preoccupied with something else (Reed 2007). The hypothesis developed by Hrdy suggests that it was deep within our evolutionary past, perhaps 1.8 million years ago, that the line leading to *Homo sapiens* began to include others in the provisioning of their young. In contrast to humans, wild chimpanzee mothers incessantly hang on to their babies and it takes at least three months before the mother chimpanzee will allow others to help (Hrdy 2005, p. 68). Orangutan mothers provide all the care for their young, and there is little contact between males and neonates (Allman 2000, p. 180). With gorillas too, the mother provides most of the care for her offspring (Allman 2000, p. 180). Since the mother is the sole caregiver in nonhuman primates, neonates are not given the opportunity to engage with others. The cooperative breeding environment of our hominin forebears suggests that non-kin as well as kin would have played a contributing role in the raising of children. Additionally, being passed from one individual to another placed selective pressures on infants to distinguish which stranger was helpful and which one was not (Hrdy 2005).

Hrdy argued that the common or natural Pleistocene family was 'kin-based, child centred, opportunistic, mobile and very, very flexible' (Hrdy 2005, p. 166). From anthropological studies, evidence suggests that the family could be made up of monogamous pairs with a number of non-breeding helpers or a male with a number of females or a female with a number of males. For instance, the Yanomami and Bari in Venezuela believe that to have a healthy baby, it requires the semen of several men. In these types of relationships, even though paternity is not a sure thing, more fathers may assist with provisioning. For Hrdy, 'the custom may also facilitate cooperation and alliances among men (often kin) belonging to the community' (Hrdy 2005, p. 79). We will see shortly this supports the so-called stakeholder theory in that all the community members have an interest at stake in rearing the child. If it is true that our ancestors lived in bands of mostly relatives, which seems to be the general agreement among anthropologists, then it makes sense to assume that acting prosocially and helping others who were in need would have been adaptive behaviour. I will expand more on this in the next chapter.

Hunter-gatherers and cooperative breeding

Evidence from studies with current hunting and gathering peoples supports the cooperative breeding hypothesis. Now we cannot be entirely sure whether current hunting and gathering communities represent the living arrangements of our Pleistocene forebears, but it seems to be the best comparison that we presently have at our disposal. Melvin Konner disputes the claim that the !Kung rely exclusively on maternal care. It is true that the !Kung fathers are less involved than mothers; however, they are more involved with their children than Western fathers (Konner 2005, p. 23). Some evidence offered by Konner suggests that family members are involved in raising the child. For instance, although mothers responded to distressed infants significantly more than others, 50 per cent of the time they were assisted by another caregiver (Konner 2005, p. 23). The evidence supports the claims about cooperative breeding among the !Kung.

We find with the Aka hunting and gathering peoples in the southwest Central African Republic and the northern people of the Congo that non-maternal caregivers also play a significant role in assisting with childcare. In one study, it was reported that babies ranging from one month to four months were passed from caregiver to caregiver on an average of 7.3 times per hour and mothers held their infants for less than 40 per cent of the time (Hewlett 1991, p. 34). Although other caregivers play a considerable role in provisioning for the young, 'Aka fathers do more infant caregiving than fathers in any other known society' (Hewlett 1991, p. 169).

In observations with the Efe of the Democratic Republic of the Congo, researchers found that mothers are not necessarily the first to hold the newborn; rather, the baby is passed around among the women in attendance. This continues in the first few weeks of the infant's life, as contact with individuals other than the mother accounts for about 39 per cent of the time and by the time the infant is eighteen weeks old, contact with others increases to 60 per cent (Konner 2010, p. 45). The studies suggest that Efe one-year-olds are with someone all of the time and spend on average about 85 per cent of their time in direct care (Konner 2010, p. 48). From these studies, Konner concludes that 'the analysis strongly confirmed kin selection theory, showing a very large predominance of relatives in every category compared with unrelated controls matched for age, sex and availability' (Konner 2010, p. 48). In another study, the authors went so far as to say that the Efe are 'the most extreme example of alloparenting in a foraging population' (Ivey 2000, pp. 857–858). The evidence suggests that with humans, it takes an extended family to raise a child.

Stakeholder theory

As mentioned earlier, Hamilton's rule explains why genetic relatives contribute to the care of their kin's offspring. As long as the cost of caring is less than the gain received from helping behaviour in relation to genetic relatedness, then this behaviour will evolve. There could be a number of reasons why people sacrifice their time to help with child-rearing. Some of the benefits of helping include developing and practising parental skills or elevating one's social status. Perhaps there are no other options available for them or there were strong social pressures to lend a helping hand. It must have been risky to leave the safety and comfort of one's group. Knowledge of the landscape and access to food and other resources would have led people to remain in the comfort of networks of kin.

The evidence from research with hunting and gathering peoples make it clear that while kinship is an important predictor of cooperative behaviour, the evidence also indicates that it was not just closely related individuals who demonstrated helping behaviour when it came to child care. The question then is, why do non-kin assist with childcare? Biologist Gilbert Roberts says that many acts that are considered altruistic have as their beneficiary an individual whose well-being the altruist has an interest or 'stake' in (Roberts 2005, p. 902). For Roberts, since humans are interdependent, then behaving cooperatively may be in one's best interest in the long term even without reciprocation. Robert's stakeholder theory fits well with other theories of cooperation. For instance, Hamilton's kin selection theory is simply a specific case of stakeholder cooperation as there are high genetic stakes involved. So for instance, an alarm call may be favoured not just because the behaviour may be reciprocated in the future but that by alarming others, those others will survive and this would be advantageous because within a larger group, the risk of predation will be reduced.

Additionally, Roberts' stakeholder theory fits well with the emergence of language that was presented earlier in this chapter. For instance, each scavenger depended on the other for success in attaining food. If an individual's life depends on others in the group for either food or protection against predators, then it is in the individual's best interest to keep others alive. And as I have argued in this chapter, there is no need for the individual to have the cognitive ability to appreciate their stake in the group. Therefore the individual will continue to be part of the coalition because it is in their best interest to do so. In a number of different conceivable situations, it is advantageous to cooperate. It is only with language that humans have the cognitive capacity to comprehend why they act in such a way and give them the capacity for greater cooperation. I will end this chapter by suggesting how kinship alliances formed within small-scale societies.

Reciprocal exogamy

To understand how humans were able to form larger cooperative groups, it will be helpful to consider how anthropologists considered how smaller groups came into existence. For anthropologist Levi Strauss, the foundation for the organization of human society lies with reciprocal exogamy (Levis Strauss 1969). Levi Strauss argued that local groups were held together by kinship, and through the exchange of women among these kin-organized societies, kinship alliances were formed. Levi Strauss believed that reciprocity and exchange were part of human nature, and since males exchanged females as a means to construct partnerships, then marrying out should bind different groups together. The main point here is not whether it was women or men that left the natal group but the fact that there is a binding mechanism between groups.

Working from the arguments made by Levi Strauss and other anthropologists, such as Robin Fox, fellow anthropologist Bernard Chapais suggests that the social structure of ancestral humans derived from pair bonding within a chimpanzee-like society. For Chapais, human pair-bonds remodelled the existing kin group, producing unique residence patterns, genealogical kinship, affiance kinship, descent, intergroup kinship and marriage rules. As Chapais puts it, 'it is no exaggeration that pair bonding ultimately brought about a true metamorphosis of hominin society' (Chapais, p. 185). For Chapais, hostile relations between groups were alleviated by the fact that unrelated males had relations with the same female. He suggests that male affines were likely to become allies because they had a mutual interest in the same female. Once a female left her natal group, she would develop bonds with the relatives of her mate. She was then affiliated with her family as well as her 'new' family. Simultaneously belonging to two distinct groups led to what Chapais calls affinal brotherhoods (Chapais 2008, p. 251). By this he means, 'the network of social relationships comprising two units of primary agnates united by one or more pair-bonds' (Chapais 2008, p. 251). Now it may be difficult to find the 'deep structure' of human society (Chapais 2008) because of the cumulative effect and diversity of culture, but it is helpful to identify a 'centre of gravity' in the disparate social patterns of any one species, including humans (Rodseth and Novak 2006). The evidence suggests that the social pattern of all primates is for one of the sexes of the younger generation to leave the group to join a bordering group. Generally it is the males in many of the monkey species, while with our closer relatives, the chimpanzees and bonobos, it is the female (Pusey and Packer 1987).

Previous anthropologists erroneously believed that animals were generally endogamous and incestuous. Although social anthropologist Levi Strauss correctly identified such things as kinship bonds, incest taboos and exogamous marriage as the guiding principles governing the social organization of early humans, he mistakenly believed that these were social constructs that had nothing to do with biological inheritance (Levi Strauss 1969). Yet there is considerable evidence to suggest that incest avoidance is rooted in our genes (Fischer et al. 2015). According to Chapais, 'if it were not for humankind's primate legacy of incest avoidance, of which the Westermarck effect is an integral part, human societies would be free of any incest taboo because culture would have had no biological substrate to work upon' (Chapais 2009, p. 82). It is not being suggested that humans did not reinforce incest avoidance by culturally imposed sanctions and taboos, but the evidence suggests that incest avoidance is prevalent in nature.

Incest avoidance

From the negative effects of inbreeding, one would surmise that incest avoidance would be found in all sexual-reproducing species (Fischer et al. 2015). The evidence does support this hypothesis because incest avoidance is widespread in the animal kingdom. Despite the fact that there is a drive to mate with one's own species, insects such as bumblebees, bees, termites and ants avoid mating with close kin because of kin recognition by pheromones (Whitehorn et al. 2009). There is evidence to suggest that even among non-social organisms, such as the tropical butterfly *Bicyclus anynana*, females prefer to mate with nonrelatives (Fischer et al. 2015). Since there is no evidence that pheromones are involved in the mating behaviour of primates, including humans, there is a need for an alternative explanation for the avoidance of incest in human populations.

Edward Westermarck speculated that siblings would avoid incest if there was close contact during early childhood (Westermarck 1891). Much of the evidence to support this theory comes from studies involving non-siblings who were raised together. For instance, Arthur Wolf studied arranged marriages in northern Taiwan and found that those couples who had lived together as young children had a lower fertility rate and an elevated rate of divorce and adultery (Wolf 1966; 1995). In other studies involving peers raised together in Israeli Kibbutzim, researchers found that kibbutz children did not marry each other and there was no evidence of sexual activity involving cohorts (Shepher 1983). In a rare study that focused on siblings rather than non-siblings, the researchers found 'early sustained cohabitation between siblings operates as a barrier specific to potentially reproductive acts rather that as a general suppressor of sexual interest' (Bevc et al. 2000, p. 159). It appears then that the critical variable for incest avoidance in humans is early consistent cohabitation. Whether animals refrain from incest due to chemical cues or from early cohabitation, the evidence is clear that incest avoidance is prevalent in the animal kingdom.

Pair-bonds

So we find that the incest taboo arises from biology and is not strictly a social construct, and it is not the case that humans alone sought mates outside their natal group; rather, primatological studies consistently show that in every sexually reproducing species, the members of at least one sex leave the natal group and breed elsewhere (Rodseth and Novak 2006). Furthermore, as mentioned above, evidence suggests that primates can identify and do demonstrate a preference for kin. What is the difference in the social patterns between humans and nonhuman primates? There are at least two main differences.

First, nonhuman primates form promiscuous multi-male and multi-female groups, while humans primarily have a network of monogamous families. This is not to say that monogamy was a smooth transition or that every human group was entirely monogamous, but as Bernard Chapais points out, for the most part, 'hominins went from promiscuity to generalized polygyny and from there to generalized monogamy' (Chapais 2009, p. 172). It is true that anthropologists have shown marital unions to be quite diverse across cultures (Gough 1971); however, it is equally important to keep in mind that 'in no human society is sexual promiscuity the sole or the main form of mating arrangement' (Chapais 2008, p. 161). Furthermore, an important finding is that 'there is no route out of monogamy; once a species has opted for this state, it seems that it can never escape from it' (Dunbar 2016, p. 51). The second difference between humans and nonhuman primates is, unlike humans, primates are cut off from the relatives they left in their natal group. 'Affinity per se, then is not uniquely human but affinity between groups, in contrast is evidently unknown in nonhuman primates' (Rodseth and Novak 2006, p. 236).

Dunbar's numbers

Recall that Dunbar suggests that our social networks consist of a number of hierarchically inclusive layers. Our personal social network is made up of 150 others, with a second layer of 500 acquaintances, and a third of 1,500 people that we could recognize. If we consider Dunbar's number of 150 and the suggestion that the minimum number needed to maintain genetic exchange without the risk of inbreeding is 500, then people would have avoided marrying their own by marrying people outside that second layer. Furthermore, recall that kin recognition is well established in the animal kingdom, and with chimpanzees and bonobos we find a preferential relationship between mother and daughter as well as mother and son (Goodall 1986). Paternity recognition, on the other hand, has not been observed in chimpanzees or bonobos.

Additionally, it is only humans who can label kinship by virtue of our possession of language. The ability to label not just biological relatives but the extended family of father-in-law, mother-in-law, brother-in-law or sister-in-law made possible the extension of social interlocking families. These kin-in-law alliances escalate the size of small-scale societies. Some anthropologists claim that kinship naming systems have nothing to do with biology, but there is evidence to suggest that people treat others as though they were biological kin. The reason for this is that 'they share with us a common genetic interest in the next generation' (Dunbar 2016, p. 273). This is an important point mentioned by Dunbar because we generally assume genetic relatedness is a reflection of the past but the real issue is the stake one has in future offspring. In-laws would have had the same incentives to help the offspring of the in-law alliance as they would with any other biological relatives.

Granted, kinship is not the only vehicle for human affinity and allegiance; however, it is significant, and historically it was often associated with the others. Anthropologists generally agree that foragers lived in groups consisting of families, typically called clans, bands or local groups. These local groups were part of a larger affiliation known as the tribe or regional group. Typically this latter group consisted of around 500 people, and since the vast majority of marriages took place within this group, there would have been a clear distinction between those within the tribe and those considered outsiders. Historian Azar Gat points out that the evolutionary logic suggesting abstractly that an individual would be willing to risk his or her life for two siblings or eight cousins can be extended to '32 second cousins, 128 third cousins and 512 fourth cousins which is, in fact, what a regional group was' (Gat 2013, p. 31). According to Gat, this dense network of close kinship through marriage is a major reason why members of the in-group tended to prefer each other over those considered as outsiders (Gat 2013, p. 31).

In addition to the regional group sharing genes, they also shared a common language and ways of doing things. The vast majority of human existence took place in this type of setting, so a shared intentionality would have been encoded in our genes much like the way we favour our kin. In Chapter 5, I will provide evidence suggesting children and teenagers have a nonspecific preference to collaborate with their group members. Interestingly, this attachment to the group is arbitrary. If humans can so easily become committed to the forces of the group, then with other boundary markers such as language, culture and a myth of common descent, commitment becomes even stronger. From all this evidence, it seems quite clear that humans naturally cooperate with their 'own'. I argue, however, that throughout history, humans have had to adapt to the changed environment and, at times, have had to overcome these natural predispositions in order for 'thick cooperation' to emerge.

According to Azar Gat, these kin tribal subdivisions present a remarkably similar pattern of social structure throughout the history of our species. He suggests that loyalty was 'extended above all to family and clan, but clans, fraternities of clans, and whole tribes habitually came together in alliances to counter external threats' (Gat 2013, p. 47). It was the language of kinship and ancestry that was appealed to in order to garner support from neighbouring tribes. This same formula applies to nations: 'outside pressure brought together and galvanized disparate and often conflicting communities that became conscious of greater kin-culture similarities only in the presence of an alien other' (Gat 2013, p. 49).

Summary

The central argument in this chapter has been that human cooperation has deep roots in the mammalian brain. Humans are predisposed to take care of their young, and this proclivity has been extended to include non-kin. I argued that nonhuman animal cooperation is essentially collaborative, and this form of cooperation is a product of biological evolution. The environmental pressures on emerging humans, on the other hand, resulted in humans having the capacity for oral communication. This ability gave them the capacity to extend cooperation beyond kin and coalition partners to include strangers. Since 'thin cooperation' has its roots in the hominin line, it seems unnecessary to argue, as the BGP maintain, that there is a need for a new mechanism for cooperation in unrelated populations. Humans naturally cooperate, and since there is evidence of trade over 100,000 years ago, well before any evidence of a belief in Big Gods, then this suggests that people were dealing with strangers without the threat of divine punishment. While the BGP suggest that Big Gods caused populations to scale up to the civilizational level, evidence suggests that it was a combination of coercion and the creation of laws and trade.

Political scientists Douglas North, John Wallis and Barry Weingast argue that the social order throughout the past 10,000 years has been based essentially on domination and governed by a leader and a network of his or her elite. For these authors, the elites managed the problem of violence by forming coalitions that limit access to valuable resources, such as land and labour and access to and control of such things as trade, education and religion to the ruling elite. They call this social structure 'the natural state' (North, Wallis and Weingast 2009). As long as the elites pass on the fruits of cooperation to those who have the potential for violence, the natural state will be stable. However, for 'thick cooperation' to emerge, there is a need to extend equality and autonomy to each individual and not just the members within the elite. This type of cooperation emerged relatively recently and is an ongoing process. I will begin the next chapter however by considering the evidence for reciprocity in the natural world. I will argue that while both direct and indirect reciprocities were important for the evolution and preservation of cooperation in humans, they are unimportant in maintaining cooperation in other animals (Clutton-Brock 2009).

4

From 'thin' to 'thick cooperation'

As has been argued, behaviour is considered cooperative if it provides a benefit to another individual and if that behaviour evolved to a certain degree because of this benefit. This seems to run counter to the evolutionary theory because it would decrease the relative fitness of the individual performing that behaviour and therefore it would be selected against (Hamilton 1964). However, as mentioned in the previous chapter, cooperative behaviour can be explained by not only direct fitness benefits but also indirect fitness benefits, as long as the behaviour is aimed at other individuals who carry the gene for cooperation (Dawkins 1989). The most common way in which indirect benefits occur is by helping relatives, and this was termed kin selection (Hamilton 1964). By assisting with the reproductive success of relatives, individuals are indirectly passing their genes on to the next generation. What is important to keep in mind here is that it is the average consequences of the type of behaviour and not the consequences of every single act (West, Griffin and Gardner 2007a). Later in this chapter, I will argue that if evolutionary mechanisms explain only the average consequences of a behaviour, then it is quite normal in nature to find apparent 'irrational' behaviour where the cooperative act affords no direct or indirect fitness benefit (West, El Mouden and Gardner 2011).

Furthermore, in the last chapter, I argued that hominins lived in groups of mostly relatives. According to Hamilton's rule, there will be an increase in cooperation when relatedness or the benefits to the recipient are higher and the costs to the benefactor are lower. This would suggest that cooperation in the hominin line most likely evolved because of direct and indirect fitness benefits. It was further argued that due to the environmental conditions faced by early hominins, there were selective pressures for what was termed a shared intentionality (Tomasello 2014). It was this greater cognitive capacity that paved the way for 'thin cooperation' among humans. It is not the case that human cooperation is special because humans were able to extend cooperation to nonrelatives as there is some evidence to suggest that cooperation is maintained between nonrelatives in birds and mammals (Clutton-Brock 2009). What is special about human cooperation is that humans can evaluate the costs and benefits of cooperative behaviour and adapt their behaviour accordingly (West et al. 2007b).

According to some theorists, the striking feature of evolution is that it has given rise to cooperation between and among individuals in a competitive world. They see cooperation as so instrumental in the evolutionary process that they suggest that 'natural cooperation' is a third fundamental principle of evolution, next to mutation and selection (Nowak 2013). In this chapter, I argue that the results from game theory models, as well as the evidence from animal studies, support my claim that animals naturally collaborate, while humans naturally engage in 'thin cooperation'. While Nowak argues that cooperation is natural, I suggest that there is a clear distinction between humans and nonhumans, and this difference must be made clear for a coherent theory of cooperation. I will also argue that while humans naturally engage in 'thin cooperation', the social order throughout the past 10,000 years was the result of this type of cooperation, as well as coercion. Finally I will show that it was only quite recently that humans developed the social structures necessary to attain 'thick cooperation'.

Prisoner's dilemma

Game theory has contributed significantly to our understanding of cooperation. The person credited with developing game theory is John von Neumann, a consultant for the Rand Corporation, and it was two employees of the same company, Merril Flood and Melvin Dresher, who developed the game known as the Prisoner's Dilemma (PD). It is aptly named because it typically is played by imagining two prisoners arrested for a serious crime, and in order to get a confession, both are offered a deal. In real life, this type of situation shouldn't happen, as the police are prohibited from giving any type of inducement to prisoners in order to illicit statements. Although there are a number of versions of the game, typically two players are paired against each other. Neither of the two knows each other and will not meet again in any future interactions. Cooperating in this game means cooperating with each other and not with the police. The rules of the game state that if player A defects and gives evidence against B, while B chooses his right to silence, A will get 'rewarded' with a lighter sentence of one year, while her rights-conscious partner will serve four years.

If both remain silent and cooperate with each other, then both will receive an intermediate sentence of two years. If both A and B decide to defect and incriminate the other, then both will be sentenced to three years in jail.

The different versions of the game all come with what has been termed a 'payoff matrix' (Nowak 2011, p. 6). If you are a 'rational' person, in the sense that you are exclusively concerned with obtaining the best outcome for yourself, you would have to consider all the possible consequences of your decision. For instance, you would surmise, if I cooperate with my partner and my partner does as well, I will go to jail for two years. On the other hand, if I cooperate and he doesn't, I am looking at four years behind bars. If I defect and he cooperates, then I get one year, while if we both defect, then I will receive three years. From this matrix, the third option is the best, followed by the first, then the fourth and the worst choice is the second option.

A person would then consider the choices faced by her partner. The partner can either defect or cooperate, and if the partner indeed defected, then the person should do so as well because if she doesn't, she faces the worst outcome on the matrix. If her partner cooperates, then she should again defect as this would result in the shortest sentence possible. Therefore, regardless of what her partner does, the better option is to defect. When PD is played only once between players, who are not known to each other, then the Nash equilibrium, that is, the optimal response to the strategies adopted by the other player, is to defect. Now if your partner was thinking the same way as you were, he would defect as well, which leaves both of you with a worse strategy than if you had both cooperated. The dilemma is that if both partners simply trusted each other and cooperated, they would have received a better deal than by acting selfishly. This means that mutual cooperation is the better option than mutual defection, but for the individual player there is considerable attraction to defect.

Reciprocal altruism

Robert Trivers in his landmark paper, *The Evolution of Reciprocal Altruism*, argued that the relationship between 'two individuals repeatedly exposed to symmetrical reciprocal situations is exactly analogous to what game theorists call the Prisoner's Dilemma' (Trivers 1971, p. 38). He reasoned that the iterated games played by two players are similar to life in the natural world as each player in the game, just like each individual in life, responds to the behaviour of the other. Trivers said that parental care can be understood in terms of kin selection (Hamilton 1964) but that his theory could account for altruistic behaviour between non-kin as well. He defined altruism as 'behaviour that benefits another organism not closely related, while being apparently detrimental to the organism performing the behaviour' (Trivers 1971, p. 39). He defined benefit and detriment in terms of the contribution to inclusive fitness (Trivers 1971, p. 39). The problem with using altruism in such a way is that reciprocity is not necessarily altruistic because it provides a direct fitness advantage to cooperating. An individual's behaviour in many cases that is called altruistic may be costly in the short term but over the long term there will be benefits. Rather than altruism, this type of behaviour is more appropriately termed mutually beneficial behaviour (West, El Mouder and Gardner 2011). Furthermore, it will be argued that mutually beneficial behaviour, in the sense that the individual understands that they will receive something in return, is found only with humans.

Nevertheless, according to the model presented by Trivers, the human cooperative system is unstable. Because of the precarious nature of the system, where a certain degree of cheating is adaptive, Trivers maintains that natural selection will favour a complex psychological system in each individual governing their own penchant for cheating and one's responses to these tendencies in others. He goes on to say that once positive prosocial emotions have evolved, this then puts altruists in a vulnerable position in that they are vulnerable to cheating, so there will be pressure for a protective mechanism to counter the cheaters. He surmised that evolution would select for the ability to detect cheating, an ability that has been termed the cheater detection device (Cosmides and Tooby 1992). Trivers accounts for a number of emotions that may have been selected for in a similar fashion. For instance, the emotion of gratitude was selected to regulate the human response to cooperative acts. The emotion of guilt, on the other hand, evolved as a successful way for repairing relationships once the guilty person's cheating has been exposed. While emotions played a pivotal role in human cooperation according to the theory proposed by Trivers, there has been little empirical evidence on the role of emotions in reciprocal altruism research. In contrast, there is a great deal of documentation on how cooperation may have evolved as a stable evolutionary strategy in formulating behaviour that is mutually beneficial and is contingent on the behaviour of others. I will return to the role of emotions later in this chapter, but first I will look at the evidence for cooperation in animals and the various mathematical models that could account for the form of cooperation that I have identified as collaboration.

Animal altruism

While Trivers coined the term *reciprocal altruism*, others prefer to describe it as direct reciprocity (Nowak 2011), but I have suggested that reciprocal cooperation or simply reciprocity is the best descriptor. What this suggests is that cooperative behaviour can materialize from the selfish motives of so-called rational individuals, and Trivers was the first to suggest that cooperation could emerge in animals in this fashion. In the last chapter, I argued that there is considerable evidence to support Hamilton's inclusive fitness hypothesis to the effect that animals prefer to be with and will assist close relatives, more so than remote relatives or non-kin (Clutten-Brock 2009). The research on reciprocity will provide a clearer picture of the difference between collaboration and 'thin cooperation'.

Altruistic or manipulative

Tim Clutten-Brock, a professor of ecology and evolutionary biology at Cambridge University, questions whether cooperation between non-kin in the animal kingdom is as common as some theorists suggest and whether the cooperative interaction between animals is comparable to the models provided by game theorists (Clutten-Brock 2009). Clutten-Brock suggests that animal behaviour may be more appropriately described as mutualistic interactions and manipulative behaviour rather than reciprocity. The logic behind this reasoning is that with reciprocity models there must be a time lag between the cost to the animal and the future benefit it will gain, and this time interval invites the possibility of cheating. Clutten-Brock is not convinced that when it comes to animal behaviour, such a time lag exists. With mutualistic or manipulative behaviour, on the other hand, the cheating possibility is excluded.

In order for us to conclude that cooperation is maintained due to direct reciprocity, Clutten-Brock says four principles must be met. First, the cooperative interactions between the animals must be done by the same individuals on a regular basis. Second, the regularity with which animals offer assistance must be dependent on the regularity with which they receive assistance. Third, cooperative behaviour results in a temporary cost to the cooperator's fitness, and the behaviour is adapted to provide benefits to their conspecifics. Finally, the individuals must not be kin or a potential mate. By considering some of the studies that have been offered as evidence to support direct reciprocity in animals, we will find that a more parsimonious explanation is that cooperative behaviour in animals is sustained by immediate shared benefits, manipulation or kin selection (Clutten-Brock 2009, p. 54).

Sharing blood

One commonly cited study to support direct reciprocity comes from Gerald Wilkinson of the University of Maryland. Wilkinson studied vampire bats and found that some of the adult bats and about 30 per cent of the juveniles return to the roost without any food. Obviously without food the bats would die. What Wilkinson found, however, was that bats rarely starve; rather, their mates would regurgitate some of the blood they had acquired and share it with the unlucky bats. Wilkinson also found that bats were more likely to share with a bat that had fed them on a previous occasion (Wilkinson 1984). Wilkinson does say that much of the roost is made up of kin so that his results do not exclude the possibility that kin selection played the primary role in the cooperation. Other alternative views suggest that perhaps manipulation was the reason for the sharing. From the evidence in the original study, it seems reasonable to conclude that the situation could be described as a form of tolerated theft in that the bat would be better off to 'share' some of its meal rather than being constantly harassed. In a follow-up experiment, Wilkinson did a two-year study with bats and found that 64 per cent of the pairs sharing food were unrelated, suggesting that kin selection was not a significant factor in the cooperation among the bats (Wilkinson 2013). While the second study ruled out kin selection, it did not exclude manipulation.

There could be another explanation for the apparent reciprocity of the bats. West and colleagues point out that the evolutionary mechanisms only explain the average consequences of the behaviour (West, El Mouden and Gardner 2011). They give as an example the helping behaviour of a dolphin that assists an exhausted swimmer. This type of behaviour has no direct or indirect benefit for the dolphin so there must be another explanation or it would not have evolved. The simple answer to this dilemma is explained by considering how this type of helpful behaviour would have been selected for if, on average, it provided a direct or indirect benefit. The behaviour of the dolphin may be a by-product of selection for helpful behaviour with other dolphins in the group. The point that West and colleagues are suggesting is that the maximization of fitness does not lead to the maximization of perfect fitness in every situation faced by the organism. They go on to say that the potential for such 'irrational' behaviour surfaces even before there is time for selection to 'catch-up' with the environmental change (West, El Mouden and Gardner 2011).

In another example cited by Nowak to support his claim that animal cooperation is maintained through direct reciprocity, I suggest that it provides clear evidence that the type of behaviour so described is collaboration and not reciprocity (Packer 1977; Nowak 2011). Craig Packer found that male baboons cooperate with each other for the purpose of ganging up on a baboon couple in order to snatch the female baboon for sex (Packer 1977). However, only one of the cooperating baboons gets to have sex with the female and so Nowak suggests the reason that the baboon cooperates is essentially 'because he expects the favour will be returned' (Nowak 2011). Yet, subsequent studies show that once the baboons abduct the female from her mate, the two attackers end up fighting over her, themselves (Nowak 2011). If this is the case, then it is not entirely clear whether the baboon 'expects' that the next time an opportunity like this arises, then his mate will necessarily help him. Does the baboon actually 'expect' the other baboon to hold up to his end of the bargain, or is it more likely that in a future encounter, the two will end up fighting over the female again? I don't believe that there is any convincing evidence to demonstrate that the baboon 'expects' to be treated reciprocally.

It is argued here that the examples offered by these theorists do not lead one to conclude that there were any agreements. If the baboon expects help in the future, then this points to a norm. If the other baboon fails to return a favour, is it reasonable to conclude that the baboon would feel guilty? I don't believe there is any evidence that baboons can have a sense of guilt because guilt requires the capacity to know that one has transgressed a norm. Additionally, while many of our emotional signals may have been important for displaying commitments, the emotion of guilt does not have the accompanying distinctive facial expression (Keltner and Buswell 1996). This suggests that emotions, such as guilt, require the cognitive capacity of humans, a capacity that is out-of-reach for all other animals. Most likely, it is every baboon for himself, and the baboon is not considering any future joint endeavours. It may very well appear that one of the baboons is 'agreeing' to help the other in exchange for a future favour, but it could also be argued that to have the capacity to make such a deal, the baboon must have a grasp of the future and the ability to calculate. This seems to be unattainable, not only for the baboon, but also for any nonhuman animal. The ability to barter, swap, negotiate or trade requires the mental capacity of a human being – a human being equipped with language. What this example suggests is that there is certainly collaboration among baboons, but that it is motivated by selfish interests.

It is not necessary to care for the other individual in order take turns to achieve a goal, but it does require strategic thinking and the understanding that there is an incentive to continue to collaborate. Two studies suggest that chimpanzees don't have the sophistication to be cognitively aware of the benefits of alternating favours (Brosnan et al. 2009; Yamamoto and Tanako 2009). In a fairly recent study, researchers compared the performance of children aged 3.5 and 5 years with chimpanzees on a collaborative task. In this experiment, both parties would eventually be rewarded if they were able to work jointly, but if they worked merely simultaneously they were not rewarded (Melis et al. 2016a). This experiment, as well as others, demonstrates that children, and not apes, can work collaboratively with a cooperative strategy to achieve a joint goal.

In the experiment, the children and the chimpanzees were seated at an apparatus for turn-taking tests. The participants were in pairs, and they each had to pull a rope in order to obtain a reward. However, only one of the dyads could access a reward on each trial. For the children, the reward was a single sticker, and for the chimpanzees, the reward was either a banana or a carrot. The researchers found that while some 3.5-year-olds were successful in developing a turn-taking strategy, none adopted this strategy from the beginning, and a number of the pairs could not come up with a solution. For five-year-olds, on the other hand, they adopted a cooperative strategy from the beginning and appeared to take into consideration the outcomes of previous trials. Crucially, five-year-olds explicitly planned to take turns by saying such things as, 'let's always take turns' (Melis et al. 2016a). This demonstrates that five-year-olds were committing themselves to a norm and understood that the other person was also committed to a norm. The evidence suggests that by the age of five, children have a shared intentionality when pursuing joint goals.

The chimpanzees in this experiment performed poorly as no pair developed a consistent strategy that involved turn-taking. The authors concluded that only humans and not chimpanzees have the capacity to develop mechanisms to support cooperative interaction. They wrote, 'the capacity to forgo an immediate benefit in order to balance their own desires with their partner's desires – not necessarily out of the concern for the other but merely strategically' (Melis et al. 2016a, p. 993). This study supports the claim that it is unlikely that a chimpanzee or any other nonhuman primate is motivated on the prospect of a future reward in reciprocal interactions. Although there is evidence to suggest that chimpanzees may act favourably with those who were good partners in the past, this does not suggest that chimpanzees have the cognitive ability to anticipate or expect future reciprocal action (Melis et al. 2008). When it comes to helpful behaviour, it seems that nonhuman primates help reactively and not proactively. Furthermore, the evidence that this type of proactive cooperative behaviour does not develop in children through interaction with others also supports the claim that chimpanzees do not have the cognitive capacity to engage in the reciprocal interactions claimed by some researchers.

Although there is some evidence to suggest that marmoset monkeys provided food to non-reciprocating and genetically unrelated individuals, this is quite different from the concept of barter or trade (Burkart et al. 2007). Humans can exchange different items, so this requires the cognitive capacity to assess the worth of the items to be exchanged. The two items need not be equivalent as long as each trading partner sees the transaction as a fair deal. The studies from animal researchers suggest that animals simply do not trade in the way that humans do. One form of sharing that differentiates between nearly every animal, except those species identified as cooperative breeders, including humans, is the sharing of food.

Unlike humans, nonhuman primates collect and eat essentially the same food. The one slight difference is that male chimpanzees tend to eat more meat than females. More important, for our purposes, is the fact that with nonhuman primates there is no food sharing between the adults. Humans are expected to share food yet, 'not even a hint of this complementarity is found among non-human primates' (Wrangham 2009, p. 135). Furthermore, one does not find primates giving or trading one type of food for another. In fact, chimpanzees fight over food rather than share, and the most intense fighting is over meat. For example, if a low-ranking chimpanzee happened to be the successful predator, a dominant male is bound to take the entire catch. The chimpanzee stealing the food tries to take it to a location to eat in isolation, but inevitably others scramble around attempting to take what they can. Sometimes persistent begging leads to the dominant male giving up a small piece of food to get rid of the nuisance (Wrangham 2009, p. 157). It seems that humans, or more likely hominins, began to share, trade and barter while other animals did not, and this can be attributed to the selective pressures to cooperate when scavenging, as described in the previous chapter.

In the last chapter, I also suggested that there were blood ties between groups that made for greater trust between them. Trade depends on trust, and trade with strangers is only made possible if there is a willingness to trust others. Archaeological evidence suggests that there was long-distance trading of obsidian, shells and ochre from at least 100,000 years ago (Ridley 2010, p. 92). The estuarine snail shell called Nassarius was something treasured, as it has turned up in a number of different sites. For instance, at Grottes des Pigeons near Taforalt in Morocco, a number of shells, some covered with ochre, were discovered approximately 82,000–120,000 years ago (Ridley 2010). Ochre, as well as fine-grained stones for spears, found at Blombos cave in South Africa dating about 70,000 years ago, is almost all from non-local sources (Sale 2006). Ridley points out that not only does this indicate personal ornamentation, symbolism and, perhaps, the use of money, but it also suggests some form of trade (Ridley 2010). Ridley also notes that Taforalt is 25 miles and Oued Djebanna is 125 miles from the coast where these items were found (Ridley 2010). This evidence suggests that there was a wide range of exploration by humans at this time, and most likely there must have been some type of cooperative trade taking place among different bands.

The take-home message from the animal studies considered in this chapter strongly suggests that evolution of cooperation emerges in nonhuman animals despite their lack of foresight, and this cooperation is restricted to collaboration. There is cooperation of a sort, but it is without comprehension (Dennett 2017). I will now turn to other game theory experiments to provide additional evidence to support the claim that this form of cooperation may emerge in a world of selfish individuals.

Computer tournaments

Political scientist Robert Axelrod in his book, *The Evolution of Cooperation*, sought to develop a theory to account for the emergence of cooperation in a world of egoists without a central authority (Axelrod 1984). He developed a computer tournament and invited a number of people from different disciplines to submit a strategy to be used if confronted with an iterated PD. Developing an effective strategy requires not only taking into account the characteristics of any particular strategy, but also having to consider the nature of the strategies of the individuals with whom he would be interacting. A competent strategy would also have to take into account the history of the interactions already completed. Once all the contributions were submitted, each programme was paired off with each of the other submissions to determine which of the strategies was best over all.

University of Toronto Professor Anatol Rapoport submitted the winning strategy – called Tit for Tat – that begins with a cooperative move and then simply repeats the partner's previous move. This strategy leads to an opportunity of establishing trust between the players. One problem with Tit for Tat, first pointed out by physicist Robert May, is that the game does not incorporate mistakes. This is an important point because in nature animals, including humans, make mistakes. A Tit for Tat strategy played by both players could lead to endless retaliation. In order to account for the possibility of mistakes, Martin Nowak continued with the work spearheaded by Axelrod and incorporated mistakes into the game to make it more true to life. By including mistakes in the programme, Nowak reasoned, the strategies would mimic natural selection as random mutation would enter the picture.

Nowak designed a probabilistic computer tournament where a number of different strategies would be generated at random and evaluated by 'natural selection'. To make it more lifelike, the strategies performing the best would be replicated and those strategies that performed poorly would be terminated. After thousands and thousands of generations, Nowak and his collaborators found that the 'always defect' strategy would take an early lead (Nowak 2011). However, Nowak soon discovered that over several generations, a consistent pattern emerged. The tournament would start with all defectors, then eventually move to Tit for Tat and then to a strategy coined Generous Tit for Tat. The latter involved the strategy of always reciprocating cooperation with cooperation, but when faced with a defector, periodically respond with cooperation. The most important point here is that the random act of kindness must be just that, random. The take-home message from this work by Nowak is that the strategy that tends to dominate is one incorporating a 'nice approach'.

After investigating the computer simulations, Nowak was not entirely happy with the occasional losing streak that came with the Generous Tit for Tat approach. He found a better strategy after looking at all the options faced by a player: 'If we have both cooperated in the last round, then I will cooperate once again. If we have both defected, then I will cooperate (with a certain probability). If you have cooperated and I have defected, then I will defect again. If you have defected and I have cooperated, then I will defect.' (Nowak 2011, p. 43). After discussing this observation with biologist John Krebs, Nowak came to the conclusion that this is a strategy animal behaviouralists term Win Stay, Lose Shift.

With the Win Stay, Lose Shift, we find a strategy where if doing well, the individual holds with the current choice, but switches course of action if not doing well. What is needed for this type of strategy is for him or her to simply have the capacity to monitor his or her own action. There is no requirement to remember previous encounters with specific others or to enter into contracts for future transactions. The cognitive ability required for this type of cooperation to emerge does not require a great deal of skill. In other words, for this type of cooperation to emerge, an organism does not require foresight, and that is essentially the type of cooperation we find with nonhuman animals – mutual collaboration.

Humans, on the other hand, have the ability to take into consideration the experiences and motivations of other people, have the cognitive skills for greater memory and the capacity to predict and plan for the future. In the PD games when played only once, the rational option according to the RCT is to defect. With repeated encounters in the natural world, there are other crucial components, with one of them being reputation. People can assess whether you would be more or less likely to cheat based on information from other people. In small-scale societies, direct reciprocity works well, as everyone can recognize other individuals. Richard Alexander coined the term indirect reciprocity to characterize how everyone in the community can choose to cooperate with others by assessing their status and reputation. Of course, in order for indirect reciprocity to work, one must have a capacity for communication. Humans can talk about their experiences with others and pass that information around, suggesting that reputation plays a crucial role in human societies.

Drawing on the evidence from these games, we can see that the results fit nicely with the theory adopted by Trivers in that moral emotions are adaptations for cooperation but this is limited to human cooperation. For instance, in the initial move, the first player is sympathetic to the second player and therefore cooperates on the first move. Faced with a cooperator, the second player feels good around a cooperating individual and cooperates as well. Faced with a defector, the player has the emotion of revenge and wants to punish the other person. This would be evolutionarily adaptive because if one is too nice, then there is the risk of being exploited. Finally as indicated, with humans, reputation is extremely important, and this brings up the problem of demonstrating one's commitment.

Commitments

In day-to-day interactions, our emotions assist in making trustworthy commitments. Take, for instance, the example offered by economist Robert Frank to demonstrate how emotions help to solve a problem that may arise in social interactions. Frank asks us to consider two entrepreneurs, Smith and Jones, who want to start a joint venture in a restaurant. Smith is a talented cook but not competent when it comes to keeping the books. Jones, on the other hand, has shrewd business skills and a clear understanding of how to run a small business. A problem surfaces because they both have opportunities to cheat without being detected by the other. For instance, Jones can pocket cash from the till and Smith can take kickbacks from food suppliers. If either person based their actions on rational self-interest, then the restaurant would not have started in the first place, as both would fear being cheated. Frank says that people don't use reason in such situations but make commitments guided by their emotions. Jones doesn't steal the cash for fear of shame or guilt and trusts Smith because Smith has a reputation of being an honest person. The same principles apply to Smith.

The restaurant venture rests partly on the fact that if a person cheats, then she would 'feel bad'. This emotional feeling achieves what a rational judgement of self-interest cannot, namely, to behave honestly even though there is a chance of getting away with cheating. Frank points out that it is ironic that the genuine long-term advantage is the result of a refusal to pursue short-term gains. It is accepted that following through with the commitment may involve 'losses', but the problem is that without being able to make credible commitments, it will be even more costly in the long term.

Acting honestly and getting along with others are valuable because they make possible other opportunities in different situations. Consider travelling to a foreign country and dining in a restaurant – a restaurant that you will never return to again. Do you leave a tip? The rational person from the point of view of RCT would not, as there is no rational incentive to leave a tip. Yet most of us, I would argue, naturally leave a tip, and this is because we are 'victims' of our emotions that evolved for a different reason, namely, to demonstrate that we are trustworthy persons.

Two very important aspects of Frank's theory should be stressed here. First, when an individual is faced with a dilemma, a judgement can act as a personal commitment. A person may feel bad if they do something wrong, and this feeling will inhibit them from making such a decision. Second, we also have to consider the public nature of the act in that whenever a person makes a judgement and behaves in a certain way, it is framed from a social perspective. In the last chapter it was argued that once equipped with language, one's commitment is not simply an internal intentional state but is also expressed publicly. With Frank's model, we can see that there would have been a fitness benefit to behave morally.

While Frank made the case that having a conscience with emotions was evolutionarily adaptive, anthropologist Christopher Boehm provides evidence to support this hypothesis. Boehm says that what keeps us from getting into trouble with others is what he refers to as Darwin's inner voice. He uses this term in reference to Darwin's claim that 'any animal whatever, endowed with well-marked social instincts, the parental and filial affections being here included, would inevitably acquire a moral sense or conscience as soon as its intellectual powers had become as well developed, or nearly as well developed, as in man' (Darwin 1859). While the BGP claim that the inner voice of a punishing God is what shapes a person's prosocial choices, Boehm counters that the evolution of human conscience began by the social control by other people and that initially this was *nonmoralistic*. This occurred because groups of individuals formed coalitions to punish those in the group who were deviant or causing problems. Boehm suggests that this form of punishment could be called 'social selection' because these types of sanctions not only motivated people to behave more cooperatively but also had an effect on the gene pool (Boehm 2012). He suggests that by groups forming coalitions, it either 'completely suppresses their behaviour at the level of the phenotype or by placing them at a net genetic disadvantage' (Boehm 2012, p. 65). The most powerful free-riders have been alpha-type bullies who terrorized others and took what they pleased (Boehm 1997). For Boehm, the reason human groups have been egalitarian for tens of thousands of years is due to the fact that humans inherited the tendencies of our hominin ancestors to resent being placed in an unequal position (Boehm 1997).

The time period Boehm is referring to is around 250,000 years ago when intensive hunting was the norm (Stiner 2009). At this time, he suggests, group punishment became severe and this 'significantly affected the human gene pool because punishment reduced the fitness of deviants' (Boehm 2012, p. 149). The thesis provided by Boehm suggests that this harsh form of social selection had acted on gene pools. He goes on to say that this led to a selection pressure favouring individuals with a nonmoral conscience, thereby having greater self-control. This reasoning suggests that those individuals who are able to internalize the rules of society are more likely to succeed than those who fail to do so. Behaviour associated with psychological feeling, Boehm argues, led to reproductive success. Boehm is not specific on dates as he concedes that this type of egalitarianism may have arrived much earlier, perhaps 500,000 to 250,000 years earlier, and he says we cannot rule out such egalitarianism among *Homo erectus* (Boehm 2012, p. 163).

Boehm is not the first scholar to claim that punitive control may have affected gene pools. Robert Trivers, for instance, argued that individuals who failed to reciprocate favours may have been shunned, or worse killed, and this may have reduced the frequencies of genes for cheating (Trivers 1971). This point is important as it was those people who failed to reciprocate who were shunned or killed. Reciprocity was the process that brought about the moral sense that served both personal commitments and interpersonal commitments. Following this train of thought, anthropologist Keith Otterbein argues that moralistic aggression in the form of capital punishment against cheaters would, over time, modify human gene pools because aggressive men would have less reproductive success (Boehm 2012). Biologist Richard Alexander (1987), on the other hand, placed emphasis on the importance of a good reputation and good social status for reproductive success.

Dominant behaviour

Others have pointed out that counter-dominant behaviour is a stable strategy, and this is an ancient and evolved human behaviour (Erdal and Whiten 1994). Erdal and Whiten placed emphasis on the difference between human groups and other primates. They maintain that the evidence from studies with chimpanzees, bonobos and gorillas demonstrates that groups of primates are led by an alpha male or in the case of bonobos an alpha female. In contrast, human hunting and gathering societies are generally egalitarian, which suggests this behaviour is an ancient evolved human pattern (Erdal and Whiten 1994). The critical difference that needs to be stressed here is the willingness of apes to demonstrate subordination (de Waal 1989). Historically, striving for power has been a motivating factor, so dominant behaviour was not necessarily lost in evolution. Rather, humans achieved a balancing act through counter-dominant tendencies. Erdal and Whiten also argue that this inclination to counter-dominant behaviour offered a fitness advantage because it would be advantageous for an individual to resist being dominated by others, but in other situations it may be advantageous to dominate. It's not the case that there was a reduction in the frequencies of genes for dominance; rather, each individual comes with a host of complex motivations that includes a tendency to resist being bullied, as well as a penchant to dominate. These competing psychological tendencies would allow humans 'the ability to hold the options in mind and measure them against important aspects of the situation' (Erdal and Whiten 1994, p. 178).

The evidence suggests that humans have a penchant for domination as well as a motivation to resist it, and if necessary, will resort to violence in order to achieve their goals (Kelly 1996). This was a biological adaptation that proved advantageous in a small-group environment. This suggests that in every society there are people, particularly men, who seek to dominate others. They will also readily take risks in order to establish their dominance (Hayden 1995). In small-scale societies, these bullies were managed by small coalitions of other men (Boehm 2012). By explaining how this biological predisposition is expressed differently in the different types of social orders, we will find that 'thin cooperation' was achieved for most of recorded history through a dominant coalition who cooperated with each other in order to protect their own privileges and resources. To unpack this argument, I will show that as the population increased, a number of conditions led to the opportunity for dominant individuals to cooperate in order to gain political and economic control.

Surplus creation

The term economy is defined as 'the way people meet their basic needs, provisioning the material means of existence' (Johnson and Earle 2000, p. 22). Johnson and Earle differentiate between a subsistent economy and a political economy. The former is organized at the level of the household to meet the basic needs of food, clothing, shelter and defence. In the subsistent economy, the goal is to minimize effort to meet these basic needs. Accordingly, the only surplus created is generally limited to just enough to get by in bad times. However, as the population increases, resources are depleted and people must turn to alternative food sources or improve technology in order to increase productivity (Johnson and Earle 2000).

With low population densities, the most effective means of production is by foraging. It has been claimed that in this type of environment, foraging is the 'original affluent society' (Sahlins 1972). At low population levels, the general response to resource depletion or competition is to move elsewhere. The costs of merging with other groups to form larger political units outweigh any benefits that such a coalition has to offer. However, as more and more land is occupied and the competition for resources increases, this leads to what anthropologist Robert Carneiro called circumscription and sociologist Michael Mann called caging (Carneiro 1970; Mann 1986). This forces people to structure more organized societies in order to defend against people in neighbouring territories. For families in these circumscribed populations, the benefits of cooperating in the political economy exceed the costs.

Earle and Johnson argue that as the population increases, the subsistence economy must increase to provide for the greater number of people (Earle and

Johnson 2000). They suggest that this leads to four kinds of problems. The first type of problem is production risk. As people occupy more land, the most desirable and abundant foods are eventually depleted and therefore security decreases in times of poor crops. Faced with this type of situation, hunting and gathering communities would simply move to another location to search for food. With a sedentary lifestyle, Earle and Johnson suggest that to overcome the resource depletion problem, people either resort to community food storage or have sharing agreements between communities. This type of organization requires leadership, and this in turn provides leaders with opportunities for control.

The second problem is associated with competition over resources. In the subsistence economy, families and groups of families avoid competition by staying clear from each other. However, in circumscribed populations, rich fertile lands are more in demand, and this leads to the benefit of violent takeovers as this form of aggression outweighs the costs. This provides incentives for better defence of resources, and this too leads to opportunities for control.

The third problem relates to inadequate resources, and this necessitates the need to develop new technologies. For instance, communities in arid lands would require some type of irrigation system. This type of technology requires the cooperation of a number of individuals, and this would require a leader to organize the group, which again grants some individuals control.

Finally, the depletion of local resources and the necessity for resources that may not be in the local population would lead to a greater demand for trade. Even in communities with limited resources, the specialization in certain goods could provide for the population as long as there is some type of trade for unavailable resources. This would not only lead to trade with neighbours, but also extend to long-distance trading. This type of trading would require someone to make decisions and would also lead to opportunities for leadership and control. All four of these problems are solved by the integration of groups with leaders, and since they require greater technologies, 'the process is iterated up the spiral to the development of the modern state' (Johnson and Earle 200, p. 32).

Johnson and Earle point to the transition between local and regional groups. In local groups, an individual, often characterized as a leader of Big Man societies, leads the local group and has greater wealth than others. The main advantage for the Big Man is his control of trade between individuals in the local group as well as between neighbouring groups. The local group is often associated with the Neolithic Revolution, but what is important here is the development of social institutions and how domination is expressed in this setting.

Elite creation

The Chief or Big Man leads a coalition in control of trade. This 'creation of elites requires the social construction of social personas' (North, Wallis and Weingast 2009). While most anthropologists consider a state consisting of a population of over one hundred thousand people, North and colleagues suggest that the 'natural state' arises as societies reach around one thousand and more and 'new forms of integrated political and economic organization develop to limit violence' (North, Wallis and Weingast 2009). Rather than differentiating between local groups and regional polities, North and colleagues reasonably suggest that throughout human history there were only three social orders: foraging orders, limited access orders or natural states and open access orders. The term limited access order refers to an elite group who has exclusive access to resources. Open access orders, on the other hand, provide greater opportunities beyond the boundaries of the elite group. The last two orders developed within the last 10,000 years.

In natural states, the problem of violence is governed by a dominant coalition that secures access to such resources as land, labour and capital and also controls trade, worship and education. By limiting access to these resources and creating rents, the elite can make credible commitments to each other, and this in turn reduces violence. The elites agree to respect each other's property rights and inclusive access to resources. North and colleagues call it the natural state because for the most part of the last 10,000 years, this was the only form of securing peace and managing violence. They acknowledge that not all natural states are the same; however, the limited access order is the fundamental way of organizing society, and the most important organization in the natural state is the interconnected relationships within the dominant coalition (North, Wallis and Weingast 2009).

The natural state was fundamentally based on domination and coercion, and the leader governed the people with members of his or her family and an elite coalition (North, Wallis and Weingast 2009). The ruling coalition had a monopoly on power and did not extend any personal or property rights to the general population. By cooperating with those people who had the greater potential for violence and to share more with those coalitions, the natural state proved to be quite durable (Ober 2015). Historian Ian Morris has pointed out that there has been much development achieved under quite a number of different political organizations (Morris 2010). So, Mesopotamia in third millennium BCE, Tudor Britain and contemporary Russia under Putin would all be considered natural states even though they are very different societies (North, Wallis and Weingast 2009).

In every natural state, the dominant coalition is bound together to protect their mutual interests. The interaction between members of the coalitions leads to regulatory behaviour based on both formal and informal rules. According to North and colleagues, the origin of property rights and legal system was due to the elites protecting their own privileges (North, Wallis and Weingast 2009). However, these rules were not applied fairly; rather, it was a certain class of people who were treated the same. It was not until these types of privileges were transformed into rights that there was a possibility for 'thick cooperation'.

For most of human history, 'thin cooperation' was accomplished through obedience to coercive autocrats who claimed authority from the divine (Ober 2015). Cooperation was enforced by a centralized authority structure, and any distribution of goods was made with the purpose of dissuading conspirators from attempting to seize power. Authority came from a god-like ruler, and the people of the state were considered subjects of the ruler rather than citizens (Ober 2015).

In an earlier chapter, I criticized the evidence provided by the BGP in support of their claim that prosocial religions with their beliefs in mean, Big Gods facilitated the rise of cooperation in large groups of anonymous strangers. I questioned the evidentiary value provided by these authors and countered that there was very little evidence to support such a big claim. The claim that not just punishment, but severe, inhumane punishment was the necessary ingredient to keep a sinful humanity in line in order for large-scale cooperation to emerge is simply not supported by the historical evidence.

Rather than seeing the threat of cruel punishment as a mechanism for cooperation, it will be argued shortly that the possibility for 'thick cooperation' resulted from the concerns and challenges raised in questioning the rationale for using such barbaric forms of punishment. It seems disingenuous to account for 'thick cooperation' without considering what it took to include a vast portion of the human species. The reason women have a greater voice in society today, in most parts of the world, can be attributed to events that took place in the eighteenth century. In contrast to the claim that severe punishment is what is needed for greater cooperation, the reason we have 'thick cooperation' is a result of our ability to reason and the freedom to raise concerns about one's welfare and the welfare of others. One of the turning points in history was the drive for penal reform that ultimately gave rise to the idea of 'natural rights'. While the discussion of 'natural rights' came to a head in eighteenth-century France, the precursor for that type of discussion arose much earlier with the Greeks. The Greeks planted the seed from which 'thick cooperation' could emerge.

The BGP contend that a threat of supernatural punishment was the necessary element for early human societies to achieve cooperation. If people are not coerced by threats of punishment, the argument goes, people will act only with their self-interest in mind. In contrast, I provided evidence in this chapter and the last that humans have little difficulty forming into groups and, in fact, naturally engage with others. Humans are dependent on other humans in order to survive, so it is advantageous to cooperate. With the arguments from Tomasello, Frank and Boehm, it seems clear that humans are natural 'thin cooperators'. The problem lies with the fact that we tend to cooperate with our own and have some difficulty in extending cooperation beyond the in-group.

Part of the problem with the BGP narrative is their confusion between what is meant by cooperation and coercion. For instance, in 'Hand of God, Mind of Man: Punishment and Cognition in the Evolution of Cooperation', the authors point to a study involving Dogon women from Mali as evidence for the SPH. Ostensibly, the women 'are *obliged* to visit menstrual huts to advertise their fertility cycle and thereby reduce cuckoldry' (Bering and Johnson 2006, p. 229, *my italics*). For these authors, the supernatural component is fundamental to understanding the power of religion in achieving cooperation. The suggestion by these authors is that cooperation is attained by the power of the supernatural element.

The question here is whether this example is a form of coercion rather than cooperation. To attain 'thick cooperation', I suggest that there must be some type of agreement between the parties involved and there must be a capacity for choice. Quite recently, in some parts of the world, women were prohibited from driving. Would that be an example of the power of religion or culture to achieve cooperation? To achieve 'thick cooperation', there must be freedom to debate and have a voice in the process. Of course, no system is perfect and there will always be those who take advantage of others, but it is with the freedom to make one's own choice that is the necessary element for the type of cooperation that I have termed 'thick'. The reason women can drive in Saudi Arabia today is due to many women choosing to defy the long-standing ban on women driving. Indeed I have argued throughout this book that the only way to achieve 'thick cooperation' is through the recognition of individual autonomy and the freedom to have a voice. In Chapter 1, I argued that the Milesians offered an environment conducive to critical discussion, and this was an example of the type of climate necessary in the political realm for an open and more cooperative society. To borrow a phrase from North and colleagues, the environment where people are free to discuss issues critically is a necessary 'doorstep condition' for 'thick cooperation' (North, Wallis and Weingast 2009).

Doorstep condition

It was in Athens, however, in the fifth and fourth centuries BCE where we find other 'doorstep conditions' for an open society where people, to some extent, governed themselves. This self-governance was restricted to property-owning men. However, in Athens, we find the mechanisms that made it possible for a greater number of people to escape brutal coercion and to have the freedom to debate political issues. What we find in ancient Athens is 'organized political choice which took it outside the ranks of the well born and relatively wealthy and assigned it clearly and unapologetically to the Athenian demos as a whole' (Dunn 2005, p. 34). What was meant by the word democracy in classical Greece is the same as it is today – it is 'utility in organizing thought, facilitating argument and shaping judgment' (Dunn 2005, p. 39). The three words that form the guiding principle of democracy, 'we the people', express human equality better than any institution that has ever been developed. Two themes emerge whenever we speak of democracy. We find a struggle to escape from domination and coercion by attempting to improve the living conditions of more and more people, as well as the determination to have everyone treated with respect (Dunn 2005). This is, I suggest, how 'thick cooperation' is achieved. Throughout history the close cousins of democracy, namely, greater freedom and human rights, have accompanied it along the way.

According to Thucydides, democracy began in Greece, and the political ideal of democracy can be captured by a speech supposedly given by Pericles in the year 430 BCE (Dunn 2005). It was a speech delivered to commemorate those who gave the ultimate sacrifice for Athens in the opening year of the Peloponnesian War. Pericles spoke of a form of government that demonstrates the greatness of Athens that should be a model for others to emulate. It was founded on the principles of equality, rendering citizens to be equally free to compete for public honours regardless of personal wealth or social status (Dunn 2005, p. 26). It was not simply a political institution but a way of life that combined a sense of personal commitment to a community, as well as a voice for a greater number of people, rather than just the well born and wealthy. Equality is the governing principle of democratic rule where the voice of each individual is given as much weight as everyone else.

This type of political organization is quite distinct from the model of the 'natural state' as explained by North and colleagues (North, Wallis and Weingast 2009). Rather than being subjects of the ruler, citizens under democracy collectively had the authority to determine the course of action and establish institutional rules. By creating rules, perceived as fair, citizens are more likely to trust those rules, as well as each other. As historian Josiah Ober states, resolving conflict based on rules, rather than patronage, resulted in sustained economic growth, as well as prompted cultural productivity and innovation (Ober 2012).

With the city states in ancient Greece, we find clear evidence that 'information exchange' is the 'basic mechanism underpinning decentralized productive cooperation' (Ober 2015, p. 63). The unique capacity of humans to use reason and share ideas allows us to extend cooperation not only within the community but also among communities. The citizens of the ancient Greek cities who had a voice in their community learned from the history of their predecessors, as well as from their non-Greek neighbours (Ober 2008). It is true that the Greeks conceived their gods as a source of justice and they believed that some forms of criminal behaviour would likely result in divine wrath. However, 'there was nothing like the full-featured ideology of necessary obedience to a divine order that helped to sustain cooperation within societies predicated on a centralized royal authority' (Ober 2015, p. 60). What enabled greater cooperation in ancient Greece was the use of reason to communicate complex ideas to those within, as well as those outside, the community.

Classical Greece was an 'outlier' in that it stood out from other premodern communities with its use of public political reasoning to make decisions about the welfare of society (Murray 1990). The 'rules of the game', along with fair institutions, is what is fundamental for a cooperative society to succeed (Acemoglu and Robinson 2012). Particular aspects of culture, such as religion and national ethics, are not important in explaining how communities are getting better at getting along with each other. As I have shown, the extent to which people trust each other is fundamental to greater cooperation, but trust is an outcome of fair institutions and not 'an independent cause' (Acemoglu and Robinson 2012, p. 57).

Daron Acemoglu and James Robinson distinguish inclusive economic institutions from extractive economic institutions by pointing out that the latter are designed to extract income and wealth from the majority of the people to support and benefit a small segment of society (Acemoglu and Robinson 2012). In extractive institutions, a small elite rule, and this creates extensive inequality, leading to the potential for violence by those who believe they could benefit from the extracted wealth. Violence is restricted by powerful elites forging coalitions with other dominant elites creating incentives for cooperation among themselves. Extractive institutions were the 'natural state' of ancient societies, and it was the development of inclusive institutions that led to technological innovation, specialization and cooperation in establishing the conditions for mutually beneficial exchange. Economist Joseph Schumpeter argued that economic growth was accompanied by 'creative destruction' (Schumpeter 1942). By this term, he meant that innovative solutions were steadily developed, driving out previous techniques and creating new forms of social organization. This would not sit well with the elites in the natural state because any new economic organization would threaten the existing economic organization that benefits the elites. As Josiah Ober makes clear, creative destruction was central to the rise of the ancient Greek world. As he puts it: 'the historically distinctive Greek approach to citizenship and political order, and its role in driving specialization and continuous innovation through the establishment of civic rights, alignment of interests of large class of people who ruled and were ruled over in turn, and the free exchange of information, was the key differentiator that made the Greek efflorescence distinctive in premodern history' (Ober 2015, p. 18). Instability and violence characterize extractive institutions, while innovation, wealth and greater cooperation are the inherent features of inclusive institutions. A fear of creative destruction by those who hold power has been a major reason why there has been no sustained increase in living standards between the neolithic and the industrial revolution (Acemoglu and Robinson 2012).

The theory that people can be motivated to cooperate on a large scale and successfully achieve collective action through decentralized authority, without resorting to ideology or the threat of divine punishment, is demonstrated by the ancient Greeks. Public reasoning and the exchange of information were instrumental in allowing the Greek states to pursue common goals and produce and share public goods without the necessity of a supernatural power to coerce them to do so. Rules and institutions developed by citizens allowed many diverse people to exchange goods and ideas, and this 'drove continuous innovation and learning' (Acemoglu and Robinson 2012, p. 102). Rather than communities 'out-competing' other communities, with the ancient Greeks we find an atmosphere where the Greek communities traded and learned from each other. It was successful technologies or ideas that 'outcompeted' each other and spread through the Greek world. They accomplished this by creating institutions that were fair. 'Athens took the first steps on the road to being a state governed not only by rules, but fair rules' (Acemoglu and Robinson 2012, p. 151).

While the argument here is that we should consider fairness when considering what laws to create or in making claims about justice, it is not entirely clear whether humans are innately fair or not. In the last chapter, I argued that humans have evolved skills and motivations for collaborating with others. This suggests that humans have an interest in equity as cooperation is more lasting if individuals share in their collaborative effort, as well as sharing in the fruits of the act. There are a number of studies to suggest that when affected personally, humans demonstrate a strong sense of fairness (Fehr and Fischbacher 2003; Gintis et al. 2003). There are other studies suggesting that humans will punish others at a cost to themselves even if the unfair behaviour is directed at someone else (Fehr and Gachter 2002). This type of behaviour would have been evolutionary adaptive in the type of environment of the hominins described in the last chapter. However, humans are discriminate when it comes to assessing fairness. For instance, for much of our history, humans have been excessively unfair to minorities, women, homosexuals and emotionally disturbed people, so this makes the claim that we have a fairness instinct questionable to say the least. We certainly do have the capacity to be fair and we unquestionably choose who we believe deserve to be treated fairly, but does this mean that humans are innately fair?

Fairness instinct

In the *Fairness Instinct*, Lixing Sun sees fairness as governing all aspects of human societies and suggests it probably accounts for the moulding of human societies that has directed the course of civilization (Sun 2013, p. 17). He goes on to suggest that not only does fairness uphold our sense of justice but fairness has also been biologically rooted in our genes. He argues that it has become abundantly clear that 'fairness in both animals and humans is an adaptive trait that evolved for settling conflicts of interest, the inevitable consequence of social living' (Sun 2013, p. 41). People tend to be strikingly upset by unfair behaviour, and the concept of fairness often is associated with a conflict between a person's self-interest and the common good. While Sun focused on human fairness, others have done some fascinating research with primates and have come to the conclusion that the root of fairness has a deep evolutionary history that can be seen in our simian cousins. While the research is alluring, the conclusion that we have a fairness instinct is not persuasive.

Primatologists Sarah Brosnan and Frans de Waal have done exceptional research on primate behaviour, and in one study with capuchin monkeys, they argue that it was the first study of its kind to demonstrate a negative reaction by a monkey to an inter-individual outcome contrast (Brosnan and de Waal 2003). The experiment involved two monkeys, each separated by a glass partition so they both could see each other. The monkeys were enticed to play a game with the experimenter, where the monkey could trade a token in exchange for a piece of cucumber. Both monkeys were quite content with their cucumber until one of the monkeys received preferential treatment and was handed a grape. When the monkey who received a cucumber for her work saw the other monkey receive a grape, she became infuriated and threw the cucumber back at the experimenter and banged the ground, apparently protesting the unfairness of the situation. For these researchers, the monkeys exhibit inequity aversion (IA), and this evolutionary adaptive trait allowed for the development of a complete sense of fairness in humans (Brosnan and de Waal 2014).

Three points arise out of this experiment. First, the monkey who received the unfair treatment most likely doesn't think of the situation as one that is unfair but sees grapes on the other side of the cage and wants them. Other researchers have argued that the apparent IA may be due to frustration that was brought about by switching the monkey from the role of receiving a high-value food to the role of a witness receiving a low-value food (Silberberg et al. 2009). In fact, they suggest that there are no studies where IA has appeared in a procedure incompatible with generating a frustration effect. They also add that while there have been a number of experiments based on the original Brosnan and de Waal model-witness procedure, not all indicate an IA (Silberberg et al. 2009). Therefore, the behaviour could be described as a temper tantrum rather than a protest for unequal treatment. Furthermore, there are other studies demonstrating that monkeys react negatively when offered the cucumber when grapes are visible even if there are no other monkeys around (Brauer et al. 2006). If there is no other monkey around to make a comparison, then it is unlikely that fairness is something that is on the monkey's mind.

Second, the reaction of the monkey that appears to be protesting unfair treatment may be reacting emotionally. The monkey may be in RAGE, in the terms described earlier by Panksepp, and the monkey is simply acting out this behaviour. To judge a situation as unfair, I suggest that the behaviour of the monkey must go beyond an affective state. If humans face a similar situation, they have the ability to assess the situation and 'reflect' on it. If a human perceives an act that appears to be unfair, the person can reflect, make a judgment, and choose to act, or not act, based on one's values and priorities. Humans can discern whether the evidence justifies the position that this is unfair treatment and ask whether this type of behaviour should be allowed to happen. In the animal research to date, there is not enough evidence to suggest that animals have this capacity.

Third, even if the monkeys in the original study, and all the others studies that used the same methodology, demonstrate that the nonhuman primates have the capacity to compare the outcomes and efforts of conspecifics, that does not mean that the animals care about fairness (Brauer and Hanus 2012). The nonhuman primates simply do not have the same cognitive abilities as humans to have an idea of equity. The argument here is that an innate sense of fairness is not necessary for the type of cooperation found in the nonhuman world. Although it will be argued later in this chapter that the laws that were developed based on fairness have had a tremendously positive impact on allowing humans to achieve 'thick cooperation', it does not seem that we have an indiscriminate fairness instinct as suggested by some theorists (Sun 2013). In fact, I will spend the rest of the chapter suggesting that humans had to design fair rules in order to promote greater fairness despite our natural tendencies to favour our own group.

The theory of cooperation developed in this and previous chapters is more consistent with human nature than the theory offered by the BGP. Humans naturally engage with others and are likely to cooperate first with kin, second with friends and finally, with effort, strangers. Humans are distinct from other animals as they have the hardwired cognitive capacity for 'thin cooperation'. As I argued in Chapter 3, this type of cooperation was the result of selective pressures in the evolutionary environment of our hominin forebears. Human society scaled up to civilizations with large populations due to elites dominating the majority of the population and having exclusive political and economic control. It wasn't that humans required the threat of supernatural punishment in order to cooperate, as in certain circumstance it was beneficial to cooperate but often there was no other choice. For the remainder of this chapter, I will deal with the most important development in human history that paved the way for 'thick cooperation', namely, the idea of human rights.

The idea of human rights

Philosopher John Searle questions whether the idea of human rights is just an attractive concept that has no reasoned foundation. After all, the American Declaration of Independence declared that each person had 'certain inalienable rights' that were 'self-evident' (Hunt 2007, p. 216), and in the 1789 Declaration of the Rights of Man, the representatives of the French people declared 'the natural, inalienable and sacred rights of man' (Hunt 2007, p. 220). In contrast, Jeremy

Bentham, the father of utilitarianism, matter-of-factly stated that the concept of natural rights was 'simple nonsense'. Searle quotes Bentham as saying:

> Natural rights is simple nonsense: natural and imprescriptible rights, rhetorical nonsense – nonsense upon stilts. But this rhetorical nonsense ends in the old strain of mischievous nonsense: for immediately a list of these pretended natural rights is given, and those are so expressed as to present to view legal rights. And of these rights, whatever they are, there is not, it seems, any one of which any government can, upon any occasion whatever, abrogate the smallest particle. (Searle 2010, p. 175)

Harsh words, but Bentham was not alone in questioning the existence of rights that people enjoy or possess simply because they are human beings. The dismissal of rights-talk has been a subject of debate ever since the idea was first raised, and it even persists today, as is evident in fairly recent writings. For instance, in *After Virtue*, Alasdair MacIntyre wrote, 'there are no such things as rights, and belief in them is one with belief in witches and in unicorns' (MacIntyre 1981, p. 71) He went on to say 'every attempt to give good reasons for believing that there are such rights has failed', and he suggests that 'the eighteenth-century philosophical defenders of natural rights sometimes suggest that the assertions which state that men possess them are self-evident truths; but we know there are no self-evident truths' (MacIntyre 1981, p. 71). The two main arguments from these two philosophers are that rights must be based on a law and, second, nothing is self-evident. Both Bentham and MacIntyre are objecting to the claim that natural law or individual rights were innate and that they could only be discovered by the use of reason. In one sense, they are correct, as rights are not innate anymore than fairness is innate. Although the doubters of rights have some pertinent claims, I believe they make a fundamental mistake in their allegations.

Recall in the last chapter, I referred to John Searle's status functions. For Searle, human rights are 'deontic powers deriving from an assigned status' (Searle 2010, p. 176). He says that we can put to rest the scepticism of the rights deniers because we are not discovering that people have universal human rights; rather, when we say that everyone has the right to life, liberty and the pursuit of happiness, we are *creating* human rights. We should not say that universal rights are nonsense on stilts anymore than we should suggest that money is nonsense on stilts. The powers that humans establish can only function if they are recognized by the majority of people.

On the other hand, one may counter that the framers of the Declaration of Independence were actually claiming that King George III had infringed their *preexisting* rights, and since these rights were violated, they were justified in forming a separate government. In France as well, the deputies were not making the claim that they were creating rights; rather, the rights of the people had been ignored. The difference, however, is that the new government received approval by the guarantee of universal rights. These rights, like other status functions, such as money, are able to function because they were recognized as such. Searle makes this clear when he states that 'one way to read this last statement is not as saying you are entitled to the existence of the rights, but rather, you are entitled to the recognition of rights that already exist' (Searle 2010, p. 183).

What about the claim made by Bentham that any right, such as property rights, must be backed up by law? For Bentham, 'substantive right is the child of law; from real laws come real rights; but from imaginary laws, from "law of nature", can come only imaginary rights' (Bentham 1843, p. 523). As Amartya Sen points out however, both ethical and legal rights have 'motivational connections' (Sen 2009, p. 363). First of all, new legislation backed by legal authority, although important, is not the only motivational connection to human rights. Consider the social monitoring by not-for-profit organizations, such as Amnesty International, OXFAM and Doctors without Borders, who historically have brought to the attention of people a number of human rights violations, without any law backing them up. Second, due to better communication technology and social monitoring, these violations have been exposed to a greater number of people who in turn have called for humanitarian intervention. Just because in some circumstances the conversation about rights may have limited political power, it does not necessarily mean that the idea of human rights is ineffective.

Positive and negative rights

Searle points out that at the time the American Declaration of Independence and the Bill of Rights were drafted, the focus was on negative rights. That is, there was no need for any positive action by the state. The state was not to interfere with the individual's right for such things as the right of free speech and the right to bear arms. In contrast, the Universal Declaration of Human Rights (UDHR) includes what are termed positive rights, such as the right to a standard of living that includes food, clothing, housing, medical care and the necessary social services (UDHR Article 25). The problem with positive rights, according to Searle, is that in order to make such a claim meaningful, there must be identification of who will pay for these 'rights'. Accordingly, if there is no obligation, then there are no rights. For Searle, the drafters of the UDHR were simply suggesting that it would be a good idea if everyone could be afforded these positive rights. He concludes that the document is a 'profoundly irresponsible document because its authors did not reflect on the logical connection between universal rights and universal obligations, and they mistook socially desirable policies for basic and universal human rights' (Searle 2010, p. 185). I disagree.

I am not arguing against the distinction between positive and negative rights, as Searle's articulation of the difference is sound. Nor am I going to get into the debate whether the right to such things as clothing and housing is feasible. Nevertheless, Searle is mistaken in his claim that to take the United Nations Declaration of Human Rights literally and seriously, it has to be interpreted as an attempt to impose an obligation on everyone. For Searle, when it comes to positive rights, there are no grounds for doing so because, unlike a country, the United Nations has no authority to enforce positive rights.

The point here is that the United Nations has little authority for either positive or negative rights, so why is it a profoundly irresponsible document when it comes to positive rights but not negative rights? The Universal Declaration not only reaffirmed the negative rights adopted by the eighteenth-century framers of the Enlightenment, but also added positive rights, such as education, social security and other 'controversial' rights. There was no mechanism for enforcement for any of these rights; rather, it was simply a set of moral obligations for the world community to pursue. Granted, it was as if they were just putting these ideas out there to be discussed without attention to a mechanism for enforcement. However, by considering what was happening between the 'super powers' at the time, there would have been no possibility of achieving general agreement about these rights if accompanied with an enforcement mechanism. Therefore, rather than an irresponsible document, I suggest that it acts as a 'conversation-starter' for international discussion and reasoning about human rights.

I am using the term 'conversation-starter' to counter Dennett's catchy phrase 'conversation-stopper' because I believe he makes a fundamental error in his suggestion that we need 'conversation-stoppers'. Dennett argued that perhaps the talk of rights is nonsense upon stilts but 'good nonsense – and good only because it is on stilts, only because it happens to have the political power to keep rising above the meta-reflections – not indefinitely, but usually high enough – to reassert itself as a compelling – that is, conversation-stopping – first principle' (Dennett 1995, p. 507). For Dennett, we *need* conversation-stoppers or what he terms 'consideration-generator-squelchers' (Dennett 1995, p. 506). By this, he means that we, as human reasoners, need a method that will 'arbitrarily terminate reflections and disquisitions by our colleagues, and cut off debate independently of the specific content of current debate' (Dennett 1995, p. 506). In order that we do not suffer from endlessly philosophizing, Dennett would have us believe that when arguing with colleagues, it is necessary to have 'something that will appeal to their rationality while discouraging further reflection' (Dennett 1995, p. 506). The point here is that we do not need conversation-stoppers; rather, everything should be open to debate. A conversation-starter, on the other hand, acts as an intuition pump. An intuition pump is another catchy phrase from the enterprising mind of Dennett, and it is characteristic of an argument that may not necessarily be demonstrative but assists us in thinking more clearly about complex problems. Historically, tradition has been used as a conversation-stopper to prevent any further discussion on a number of issues and this, at times, has impeded the possibility for 'thick cooperation'. For instance, a conversation-stopper such as 'because God says so' has been used historically to prevent discussion about marginalized people. Conversation-starters, on the other hand, open debate, leading to question-generators, and this is important for expansive cooperation. I will now spend some time unpacking an argument by a contemporary theologian who calls for upholding tradition even at the expense of the rights of an individual.

Countering human rights

In *Theology in the Public Square*, Gavin D'Costa claims that humans cannot be moral without some sort of tradition to follow, and he has a bone to pick with those who see human rights as the only contemporary candidate for universal moral discourse (D'Costa 2005, p. 90). For D'Costa, those who advocate for human rights are surely wrong (D'Costa 2005). Having a distaste for anything associated with Enlightenment thinking, D'Costa says that it is only modernity's global pretensions that suggest there can be moral discourse between cultures (D'Costa 2005). He gives as an example the change in the Indian Constitution of 1950. He says that the 'great' legal historian Pandurang Kane argues against the changes in the Constitution because it was a complete break with Hindu traditional ideas. Kane complains that too many people are demanding rights when the teaching of the *dharma* was about duties and obligations. This atrocity, according to Kane, will lead people to have the right to 'impose their will' and will give them a voice for their own ideas (D'Costa 2005).

D'Costa acknowledges that Kane was a Brahmin. Brahmins are the spiritual leaders of the Indian caste system, who traditionally held considerable power in

India. The three other castes are the Kshatriyas (nobles or warriors), the Vaisyas (artisans and craft producers) and the Shudras (unskilled) (Bellah 2011). Outside the caste system were the Dalits, referred to as the untouchables, a demeaning terminology as a result of the traditional belief that they were filth. Gandhi, who was born into the Kshatriya, bravely fought for the rights of the Dalits and referred to them as Harijans (children of God).

Not surprisingly, the other castes celebrated the change, and this is acknowledged by D'Costa. However, he stresses that 'Kane's point still stands: the Indian Constitution is inimical to Hindu ethics, which is based primarily on duties, not rights' (D'Costa 2005, p. 90). D'Costa makes no mention or has little consideration for the welfare of the Harijans. Would Kane have penned this same argument if he was a Harijan? Unlikely, because, at the time of the constitutional change, Harijans were forbidden from entering temples or schools and were prohibited from receiving an education. Fortunately, since the changes in the constitution, some people whose *dharma* was to live in slums and clean toilets now have some voice in the rules of society. Of course, not everything is promising in India, or any other country for that matter, but the point is that people who were destined to be on the brink of starvation now have a trace of hope for a better life. This is an accomplishment that could not have happened without recognizing the autonomy of the individual and questioning the merits of a tradition.

D'Costa's argument rests partly on the fact that the concept of human rights is Western in origin, and he contends that the imposition of Western ideas, such as human rights, on other cultures or nations is just another form of imperialism. As Martha Nussbaum points out however, unless we can give comprehensive reasons for why other cultures should not adopt the notion of human rights, other than the claim that it is Western, then we have said little of substance (Nussbaum 2011). She reasons that non-Western nations do not resist Marxism because it is of Western origin. Adopting the philosophy of Marxism can lead to many problems, but the reason not to accept it is not because the idea was born in Britain by a German Jew. Throughout human history different cultures and nations have shared ideas, and it is argued here that the idea of human rights was an essential element to achieve 'thick cooperation' in the world.

Nussbaum also calls attention to the fact that the framers of the international human rights movement in 1948 came from a host of different nations. The leaders of the diverse nations framed the talk of rights in a fashion that would make it more palatable for people from a wide range of traditions to accept. Ironically, the United States, often considered the poster child for contemporary imperialism, has not ratified two important human rights documents, namely, the Convention on the Elimination of All Forms of Discrimination against Women (CEDAW) and the Convention on the Rights of the Child (CRC) (Nussbaum 2011, p. 104). Those nations that have failed to ratify CEDAW include the United States, Iran, Somalia, Sudan and Vatican City, while only the United States and Somalia have failed to ratify the CRC (Nussbaum 2011). The suggestion that the United States and other Western nations impose their values and force human rights on a reluctant East shows 'gross ignorance' (Nussbaum 2011). The concept of human rights and dignity is not a monopoly held by the West but an idea that goes a long way to lending a hand to the weak against the strong.

D'Costa proceeds to develop his thesis by turning to the tradition of wife-burning that took place in some parts of India. He tells the story of Roop Kanwar, an eighteen-year-old woman from Rajput. She had been married for only eight months, when tragically her husband, Maal Singh, died. At the funeral, with her husband in her arms, Kanwar took her own life by being consumed by the flames. D'Costa notes that between 1943 and 1987 at least thirty women have perished in this manner in the Rajput/Shekavati region.

D'Costa does some etymological acrobatics in attempting to avoid using the term wife-burning, as he suggests that it is an incorrect rendition of the word *sati*. For D'Costa, 'the *sati* chooses other than widowhood for she is not technically a widow until the funeral pyre is lit' (D'Costa 2005, p. 149). He adds that technically, the husband is not yet dead until his *atman* leaves the body. Incredulously, he concludes that in both circumstances, 'the *sati* is never a widow' (D'Costa 2005, p. 149). D'Costa says he partly agrees with the suggestion that the practice of *sati* is abhorrent, yet he certainly seems to be sympathetic to the practice. He says that the practice is extremely complex and that he does not believe 'that the practice can be dismantled until the full power of its religious vision is understood' (D'Costa 2005, p. 151). In contrast, putting the practice in perspective, Madhu Kishwar and Ruth Vanita reasonably ask, 'If a woman does not have the right to choose whether she wants to marry, and when, and whom, how far she wants to study, whether she wants to take a particular job or not, how is it that she suddenly gets the right to take such a major decision as whether she wants to die'? (D'Costa 2005, p. 154). D'Costa counters that this type of criticism misses the point because, he suggests, 'a woman's choice is not a right, but a matter of following one's duties' (D'Costa 2005, p. 154). This is a similar argument to the one we heard from Kane. D'Costa goes on to say that within the *dharma*, there is no place for rights, only duties. That is, according to D'Costa there is an important difference between men and women since the former have no *sati* option. He reminds the reader that within the *dharma*, there is no such thing as rights or equality between the sexes or castes, and righteousness can only be achieved by following one's own path. D'Costa sees equality as an extremely modern ideology, as if it is a bad thing, and he counters that 'intra-systematically *sati* is an optional ritual to which is ascribed considerable merit, both for the woman and through her actions, for her husband, and both their families' (D'Costa 2005, p. 154).

D'Costa suggests that one way to *understand* the defence of *sati* is from an eighteenth-century text written by Tryambakayajvan, who uses scripture and tradition to demonstrate why *sati* makes a woman truly virtuous. For D'Costa, Tryambakayajvan makes some very important points so that *sati* can be understood appropriately. One of the very important points for D'Costa is that *sati* is not suicide, as the good wife is simply following her proper *dharma*. He says it is similar to the warrior who fearlessly goes to battle, knowing he will be killed (D'Costa 2005, p. 154). The second significant point for D'Costa is that the practice is meritorious (D'Costa 2005, p. 154). He says that Tryambakayajvan puts it *clearly* when he says, 'women, who due to their wicked minds, have always despised their husbands and behaved disagreeably toward them' can perform the ritual act, 'whether they do this on their own free will, or out of anger, or even out of fear – all of them are purified of sin' (D'Costa 2005, p. 154.). D'Costa believes that anger and fear may be conceived as forced *sati*, but he goes on to claim that it is clear that Tryambakayajvan is not intending to support these forms of *sati*, for his argument is 'precisely to extol the free choice of *sati* because it is dharmically coherent and attractive in securing righteousness in the world' (D'Costa 2005, p. 155).

For D'Costa, what is striking about the practice of *sati* is its atoning power, not only for the sins of the woman but also for those of her husband and family as well. He says that 'one begins to understand why the Rajput crowd (who support *sati*) want to venerate Kanwar as a Devi' (D'Costa 2005, p. 155–156). He goes on to suggest that 'what is so startling about the intra-textual logic of his (Tryambakayajvan) exposition of the *dharma* is the positioning of *sati* as heroic virtue, whereby the freely undertaken self-sacrifice of a woman is able to atone for her sins' (D'Costa 2005, p. 156). D'Costa makes it clear that he believes that Kanwar freely chose to follow her husband, and he sees this as heroic and virtuous. He sees a prohibition against *sati* as premature and asks, 'does theological discourse end up as colonial discourse in opposing *sati* extra-textually, or is it able to reach deep into the Hindu tradition employing an analogical imagination, not affirming or criticizing it (as yet), but seeking an understanding that touches, even obliquely, the divine?' (D'Costa 2005, p. 175). D'Costa concludes the chapter by arguing that he has 'tried to show why unambiguously opposing *sati* is *problematic* from a "theological" religious studies point of view' (D'Costa 2005, p. 176, *my italics*).

The main criticism of D'Costa's objectionable defence of *sati* is that all the victims simply did not have a choice. To be a truly cooperative society, a society defined as 'thick cooperators', each individual within that society must freely choose a course of action. The choice must be theirs to make. Without providing opportunities for the individual to choose their own course of life, we risk the entrenched social injustices that we find with some 'traditions', especially those that are the consequence of discrimination and marginalization. For individuals to be given a choice, they must be seen as equal. All individuals deserve equal treatment from the laws and institutions in any society.

Furthermore, D'Costa's defence of *sati* also brings to mind Amartya Sen and Jon Elster's notion of the constraint on preferences and satisfactions. They argue that if certain things are placed out-of-reach for some citizens, they may form adaptive preferences. So a woman brought up in the tradition of the virtuous 'good wife' espoused by D'Costa may not even consider the possibility of having a voice. Women and other marginalized people who traditionally have been barred from getting an education may learn not to want such an education because it has been so entrenched in their culture that those types of preferences are out of bounds for their particular gender or class. Historically, reinforcing the status quo has resulted in many injustices. Equality has been a long struggle for many, and a brief look at the history of that fight will help us understand why it is the essential element for 'thick cooperation'.

A brief history of human rights

Political scientist Benedict Anderson coined the expression 'imagined communities' to describe the impact of print technology on society. For Anderson, the ability of individuals to read novels and newspapers led to networks of shared culture beyond the 'real' person-to-person interaction (Anderson 1983). According to Anderson, print technology had a major effect on nationalism. Historian Lynn Hunt liberally adopted this idea and suggested an 'imagined empathy' was needed in order to get the human rights revolution off the ground. By this, she means that once individuals had the opportunity to read novels, it generated greater empathy as well as an awareness that there are other people like them. Hunt points out that human rights depend on both emotion and reason, and the development of such ideas occurs through interaction with others. Once people read or talk about what they are reading with others, it results in 'a notion of a community based on autonomous, empathetic individuals, who could relate beyond their immediate families, religious affiliation or even nations to generate greater universal values' (Hunt 2007, p. 32). According to Hunt, reading novels allowed readers to exchange ideas, break down the traditional barriers between people and gave people the opportunity to put themselves in the shoes of others.

Regardless of whether empathy played a significant role in the increasing conversation about rights, the important point is the fact that the subject of rights was being discussed. Thomas Jefferson, for instance, prior to 1789, used the term 'natural rights' rather than 'rights of man', and it was not until after the French Revolution that he actually spoke about human rights (Hunt 2007). While the English kept to the 'natural rights' phraseology, it was the French who invented the term 'rights of man'. Importantly, Marquis de Condorcet defined the phrase in terms of the security of the person, property, fair treatment and the right to participate in political issues (Hunt 2007).

What was important for public discussion on natural rights was the idea of autonomy. For instance, the concept of 'Divine Right' gave way to the conception of a social contract, where individual autonomous men entered into a social contract with each other. Privilege for the ruler and his small elite, granted by birth or military power, slowly gave way to less-powerful members of society. The Enlightenment writers criticized special interest for any individual. Diderot for instance, writing in 1755, urged readers to tell themselves often, 'I have no other true, inalienable naturals rights than those of humanity' (Hunt 1996, p. 37). He concluded his short article on 'Natural Law' by suggesting that laws 'should be made for everyone, and not for one person' (Hunt 1996).

The notion of autonomy was reflected in the new legislation that was being passed at the time. For instance, for the first time in France, both male and female children had equal right to inherit wealth. The age of majority was lowered from twenty-five to twenty-one years, allowing young adults more autonomy at an earlier age, and children were no longer incarcerated without a hearing. Within the family, the father's control was cut short with legislation preventing their exclusive authority over children, and family councils were created in order to hear family disputes between children and their parents (Hunt 2007). More than any other concept, human rights transformed how individuals interact with each other, and for the past 200 years, the idea has expanded to include more and more people.

Hunt points to an incident that she considers instrumental in the human rights revolution. In 1762, a 64-year-old man named Jean Calas was convicted of the murder of his son, allegedly in order to prevent his son from converting to Catholicism. The accused was sentenced to death by the customary breaking on the wheel, but prior to death, he was subject to the 'preliminary question'. The purpose of the torture was to illicit confessions from the accused and for him to name any accomplices who may have assisted with the crime. Remarkably, Calas did not 'break' after two shots on the wheel, so the authorities tied him down, forced his mouth open and poured two pitchers of water down his throat. Calas refused to cooperate, and the 'Calas Affair' sparked moral outrage and a campaign to help the family, spearheaded by Voltaire.

In that same year, Voltaire published, 'Treatise on the occasion of the death of Jean Calas'. In the essay, Voltaire did not concern himself so much with torture, but was much more troubled by religious bigotry. For Voltaire, it wasn't about empathy; rather, he was concerned with the unreasonableness of the process. In his essay, Voltaire wrote, 'it is impossible to see, how, following this principle (human rights) one man could say to another, believe what I believe and what you cannot believe or you will die' (Hunt 2007, p. 74). The core issue for Voltaire was with the rights of the individual. Voltaire learned that the family had actually covered up the son's suicide because of the public distaste for taking one's life, and, after Voltaire wrote letters on behalf of the welfare of the wife and son of the accused, his focus slowly began to shift.

It was not just in France where attitudes were changing about the use of torture and discussions about legitimate forms of punishment, but a number of other countries in Europe gradually followed suit. Some Italian states, Swiss cantons, as well as France, offered prizes for writing competitions on penal reform. In Italy, Cesare Beccaria published 'Essay on Crimes and Punishment', where he rejected cruel punishment and even questioned the effectiveness of the death penalty. Not surprisingly, there were counterarguments against the positions held by other Enlightenment thinkers calling for penal reform. Pierre-Francois Muyart, for instance, argued that what made the traditional system work was 'precisely because each man identified with what happened to another and because he had a natural horror of pain, it was necessary to prefer, in the choice of punishments, that which was the cruelest for the body of the guilty' (Hunt 2007, p. 94). Muyart's thirst for vengeance is intuitively pleasing, and he defends his position in the same manner as the BGP. However, both overestimate the efficacy of harsh punishment.

In France, once the idea of rights was born, others pushed for the rights of others. For instance, the Protestants argued that they had the same right as Catholics to hold office. Decreeing that non-Catholics could hold office did not take very long before the notion about non-Catholics included more than just Protestants, but extending to Jews as well. While rights for Free Blacks and slaves was a long battle in the United States, it was France that led the way emancipating the slaves in 1792, well before any other country. By 1790, 'actors, executioners, Protestants, Jews, free blacks, and even poor men could be imagined as citizens by at least some substantial number of deputies' (Hunt 2007, p. 170). While political leaders could agree or be convinced into agreeing that rights applied to all men of different religions and even without distinction in the colour of a person's skin, very few were committed to the idea of rights for women.

Sophie de Grouchy and her husband Condorcet were the most outspoken supporters for the rights of women. Keeping with the logic of the rights revolution, Condorcet published a newspaper article in 1790, 'On the Admission of Women to the Rights of Citizenship', calling for equal rights for women. Citing the rationale used to encourage greater rights for men regardless of wealth, colour or religion, Condorcet reasonably argued the same should be applied to women. He questioned whether the legislators had not 'violated the principle of equality of rights by quietly depriving half of mankind of the right to participate in the formation of the laws, by excluding women from the rights of citizenship' (Hunt 1996, p. 120) After all, he reasoned, since women share the same feelings and are as capable as men in acquiring moral ideas through reasoning, then they should be afforded equal rights. He maintained that 'either no individual in mankind has true rights, or all have the same ones' (Hunt 1996, p. 120).

Others also campaigned for women's rights. Like Condorcet, Olympe de Gouges closely followed the structure of the 'Rights of Man and Citizen' to demonstrate that women were not afforded the same rights as men. In the 1789 Declaration of the Rights of Man and Citizen, the opening statement reads, 'Men are born and remain free and equal in rights. Social distinctions may be based only on common utility.' (Hunt 2007, p. 221). In The Declaration of the Rights of Woman, de Gouges opens with 'Women is born free and remains equal to man in rights. Social distinctions may be based only on common utility.' (Hunt 1996, p. 125). Lobbying for equality comes with costs, and de Gouges' agitation for the rights of women cost her her life.

Mary Wollstonecraft was another prominent campaigner for the rights of all disadvantaged people. For instance, she was critical of others, such as Edmund Burke, who cherished the rights and liberties of some individuals in society while ignoring the predicament of others (Wollstonecraft 1792/1995). Her criticism rested with the fact that Burke focussed his attention on defending the freedom of non-slaves but failed to address the rights of slaves. The main point of her argument is that it is inconceivable to defend the freedom and rights of only a favoured group. As indicated earlier in this chapter, the same criticism can be directed at those who view certain types of people as 'untouchables'.

Summary

The evidence in this chapter suggests that cooperation in the form of collaboration can evolve as a stable evolutionary strategy. It was argued that the difference between humans and nonhuman animals is that humans have the greater cognitive capacity to comprehend agreements and are motivated by future rewards in reciprocal exchanges. This suggests that with the capacity for greater understanding, humans naturally engage in 'thin cooperation', while cooperation for animals is essentially collaborative. In addition to 'thin cooperation', humans also have the ability to extend cooperation beyond their in-group, but this can only happen with the recognition that each individual is an autonomous agent, free to determine the course of their own lives. In the next chapter, we will find evidence from developmental psychology that supports the claim that humans naturally engage in 'thin cooperation' and that children reason about their behaviour when interacting with their peers.

5

How does that make sense?

A number of years ago, when my youngest son was five years of age, he said, 'you know, you are never really doing nothing, because even when you're doing nothing, you are always doing something, watch!' At which point, he set out to prove his hypothesis by sitting on the step with his hands on his lap remaining still for a few seconds. He wrapped up his demonstration by declaring matter-of-factly, 'see, I'm not doing anything, yet at the same time, I am doing something, I'm sitting'. He reasonably inquired, 'how does that make sense?'

Curtis was not attempting to be smart or sarcastic; rather, he was reasoning about something on his mind that he found paradoxical. Part of the purpose of this chapter is to provide evidence to demonstrate that children contemplate issues and reason much more than they are generally credited for and they attempt to figure out, for themselves, how things make sense. This is especially true when children are engaged in social interactions. Psychologist Jean Piaget convincingly argued that children develop ideas about morality through interaction with peers and do not necessarily conform to the norms set by adults or other authority figures. According to Piaget, children must learn how to deal with problems, how to treat others with respect, and this can only occur through social interaction with their peers. For Piaget, morality must not be defined by the norms or rules of a particular culture but rather from the principles deriving from reasoning about how we treat other people. By resolving conflict through negotiation and argumentation, children construct concepts of equality, justice and fairness and create for themselves a personal moral understanding.

This chapter will take shape in the following manner. From the studies outlined in this chapter, we will find that unlike our simian cousins, the majority of children help and share with others. Furthermore, unlike apes, children are proactive in their helping behaviour. That is, they do not need to be cued to help or be offered rewards or threatened with punishment to act prosocially. Furthermore, children are more likely to help and share with others if they are given a choice to do so, rather than being told by an authority figure how to behave. It will also be argued that children distinguish between moral principles and conventional rules, and the vast majority of children find moral transgressions, such as assault, theft or murder, wrong regardless of whether God proclaims it wrong. This chapter provides evidence to suggest that children have a natural predisposition for 'thin cooperation'.

To begin, one question to consider is, at what age do children have the psychological structures necessary for reasoning about one's own actions and intentions, as well as the behaviour of others? Studies by developmental psychologists suggest that it is not until the preschool years that children consciously cooperate with other children and make choices in pursuit of a common goal (Hamann et al. 2011). If an individual has in mind the intention to help another, then it would follow that the capacity to accurately represent the minds of others is essential for early prosocial behaviour. In an earlier chapter, I argued that humans are capable of making automatic, unconscious inferences about the beliefs, desires, motivations and intentions of others at a fairly early age (Gopnik 1999). As a child develops this ability to reflect consciously on oneself as an object of another's perception and thoughts, as well as an agent with similar thoughts and intentions, then collaborative action becomes more autonomous and flexible (Brownell 2011). Psychologist Vasudevi Reddy argues that while infants don't consciously share the intentions of the adult caregiver, it is in this dyadic interaction that the infant learns the essential skills for joint action and cooperation (Reddy 2008). As Reddy points out, the ability to cooperate with peers and engage in a collaborative effort towards a common goal is preceded by years of joint engagement with adults.

Reddy suggests that if we take as the starting point for understanding people as being 'not isolation and ignorance, but an emotional relation and a psychological awareness', then it would change the way we approach the question of understanding minds (Reddy 2008, p. 4). She says that the traditional view about shared and private experiences is to begin with privacy as the starting point from which we can speak about private mental states. Reddy offers convincing evidence that joint action and shared experience occur before the child can communicate those private experiences. There is a long history of sharing mental states through an engagement with each other's thoughts, feelings and intentions. For Reddy, this 'public' sharing is the essential step necessary for the 'private' concealment and not the other way around (Reddy 2008, p. 16).

William James questioned how it would be possible to survive if you were surrounded by a number of people yet there was no one who took notice of your presence (James 1890). For James, being the object of another person's attention must be the first experience that we have of mentality. Therefore, an infant's first feeling of attention is in response to receiving it from an adult, in engagement. Studies have found that if an adult withdraws attention from the child, then the child becomes upset (Reddy 2008). For instance, Ed Tronick developed the ethically questionable still-face experiments, where an adult engages in face-to-face exchanges with an infant and then, for no apparent reason, withdraws from the interaction and stares blankly at the child, not responding in any way (Reddy 2008, p. 73). This raises the question, if the infant is not sensing that the adult is aware of them, then why would the infant be so upset at such a withdrawal?

In the first year of life, the adult's attention is first felt by the infant as a response to the engagement with the adult (Reddy 2008). In the early years, the infant does not necessarily infer the intentions and desires of the other, or consciously share a joint commitment with them. During the second half of the first year, the infant's engagement with the attention of others expands to include objects and events beyond the adult and child interaction (Brownell 2011). Infants eventually initiate the joint action, and the interactions become more complex. It is not until about eighteen months of age when the infant becomes the active participant, rather than being the passive player led by the mother (Brownell 2011). It is this type of engagement that provides the foundation for the development of skills necessary for pursuing cooperative joint activity. By looking at the studies regarding the helping and sharing behaviour of children and comparing those studies with research with chimpanzees, we will find that children, through the interaction with peers, show signs of 'thin cooperation' quite early in life.

Helping behaviour

In an influential study to test whether humans are uniquely cooperative, researchers compared human infants and chimpanzees on their inclination to help others (Warneken and Tomasello 2006). The researchers believed that this comparison would provide information about the features present in the LCA from the features of helping behaviour that evolved only in the human lineage. This was the first study comparing humans and nonhuman primates in a number of different instrumental helping situations. In the first study, Tomasello and Warneken presented twenty-four 18-month-old infants with ten different situations in which the adult male experimenter required help to achieve a goal.

In all the situations, some problem occurred, such as an out-of-reach object or the adult was hindered by a physical obstacle. For each of these tasks, there was a corresponding control task, where a similar situation was arranged; however, there were no cues from the adult indicating that help was needed. The reason for the control task was to ensure helping behaviour was being captured, rather than the child simply attempting to restore order. There were three phases for each problem. For the first ten seconds, the experimenter focused his attention solely on the object. This was followed by another ten seconds, where the experimenter would shift his gaze from the object to the child and back again. For the final ten seconds, the experimenter verbalized his problem and continued to gaze back and forth between the child and the object. Results found that infants provided help in the majority of the tasks and that twenty-two of the twenty-four infants assisted in at least one of the categories. Importantly, in 84 per cent of the helping acts, the child responded within the first ten seconds, before the adult either looked at them, or made a verbal request for assistance. This suggests that it was unlikely that the helping behaviour was motivated by empathetic concern for the experimenter, as there was no eye contact between the two. Furthermore, the study made clear that the children were assessing the situation, as when it was evident that the adult did not require assistance, the infants did not offer help. The researchers concluded that by the age of eighteen months, infants willingly help a stranger complete a task without receiving a reward or threatened with punishment.

The chimpanzees, on the other hand, did not fare as well as the infants, and in order to even conduct the study, the researchers had to make some significant adjustments to get the chimpanzees to comply. The helping behaviour that the authors accredited to the chimpanzees was not with a conspecific but was with the human experimenter. The reason for this adjustment is due to the extreme competitiveness of chimpanzees and their unwillingness to cooperate with others, unless it is with chimpanzees of whom they are extremely tolerant (Hare et al. 2006). Furthermore, if the targeted resource is easily monopolized by a dominant partner, then the subordinate chimpanzee tends to back off, resulting in a failure to cooperate (Melis et al. 2016b). Studies suggest that if given the opportunity, chimpanzees will most likely monopolize the resource, especially if it is food (Hare et al. 2006). Even with this modification, there was a significant difference between a chimpanzee's and a child's motivation.

In this particular study, the chimpanzees were not rewarded for retrieving out-of-reach objects for the experimenter, as the researchers were trying to determine instrumental helping behaviour and any reward or prize would certainly undermine the study. However, in previous studies, these same chimpanzees were rewarded for helping behaviour, so this may have played an influential role in their helpful behaviour in the present experiment (Warneken and Tomasello 2006). What we can conclude from this seminal study is that chimpanzees may show signs of helping behaviour, but this type of collaboration is highly constrained by their competitive streak. Cooperation for chimpanzees is reserved for only those who are highly tolerant of each other, and the resource must be such that it cannot be easily monopolized. This experiment supports the claim that chimpanzees may be able to work together, but it seems that this type of cooperation can only be described as mutualistic collaboration. In contrast, it is clear that by eighteen months of age, human infants willingly help strangers, demonstrating the capacity for 'thin cooperation'.

In a subsequent study, the same researchers using the same experimental paradigm asked whether even younger children would carry out acts of instrumental helping (Warneken and Tomasello 2007). The authors believed that around twelve to fourteen months, there is a considerable change in children's social-cognitive development as they begin to engage not just in playful cooperative behaviour with adults, but also in behaviour that requires comprehending the interdependent roles necessary for a shared goal. With helping behaviour, the researchers found that at fourteen months, infants exhibit a proclivity to help, but it was limited to certain tasks. They conclude that 'helping behaviours ontogenetically precedes cooperative activities, corresponding to the understanding of individual intentional action versus the formation of shared intentions' (Warneken and Tomasello 2007, p. 292).

Proactive prosociality

In the above examples involving human subjects, all the children were engaged in what can be termed 'reactive pro-sociality' (Warneken 2013a). That is, the helping behaviour was prompted by obvious behavioural or communicative cues by another person who was seeking assistance. 'Proactive pro-sociality', on the other hand, is engagement in helpful behaviour without being cued by the other person and the helper reacts by assessing the situation and the needs of the other person (Warneken 2013a). For example, consider being at a busy train station and imagine that a person is pushing a baby carriage towards a set of stairs. The person pushing the carriage does not have to begin the climb up the stairs or demonstrate explicit behaviour to show to others that they are having difficulty. Most people witnessing this would be able to infer that the person intends to climb the stairs. They would be cognitively sophisticated enough to think that it may be difficult to ascend the stairs with a carriage and that the person would most likely appreciate some help. The onlooker would not need an invitation in order to offer help. Similarly, there are a number of studies suggesting that children, as young as two years of age, naturally help others and, at times, they spontaneously help, even in situations when not cued to do so (Warneken 2013a). In previous chapters, I suggested that this type of behaviour is a special form of cooperation and that it is unique to humans due to our particular evolutionary history. In the following studies, we will find evidence to support the claim that young children assess the contextual cues and the behaviours of others to determine whether another is in need of help.

Researcher Felix Warneken chose children between the ages of twenty-one and thirty months. The reason he chose this age range is due to the fact that in previous studies, researchers had found that the development of helping behaviour towards unfamiliar individuals first appears between the ages of eighteen and twenty-one months (Warnekan and Tomasello 2006; 2008). Additionally, at the age of thirty months, studies indicate helping behaviour in children is very strong (Svetlova 2010). This particular study focused on children under thirty months of age, and it included two conditions (Warneken 2013a). There was an experimental condition where the child observed an empty can roll off the table onto the floor without the adult realizing what had happened. In the control condition the adult intentionally discarded the can. The purpose of this experiment was to determine whether children who engaged in proactive prosociality would either pick up the can and hand it to the adult or put the can back on the table. The researchers also wanted to see whether the child would verbally alert the adult or would the child be more likely to use a pointing gesture. The reason why it was important to determine whether the child communicated with the adult is to confirm that the child was not simply blindly picking up the can but was actually exhibiting helpful behaviour to a person who appeared to be in need of help.

The results were compiled using three groups of children aged 22, 25 and 28 months. When the researchers included cases of informing, 33, 67 and 83 per cent, respectively, of the children per age group exhibited at least one proactive prosocial helping act during the experimental condition. This study clearly shows that children at around the age of two years engage in proactive helping without being requested or cued to assist a person in need. Furthermore, this motivation to assist a stranger was strong enough to cause a child to give up a playful, enjoyable activity in order to assist another person. Most importantly, since motivation to help increases with age, this suggests that as children develop social-cognitive skills, they have a greater proclivity to engage in helping behaviour.

Selective helping

The above studies suggest that children as young as two years of age demonstrate helping behaviour, but the question now to be asked is, are children indiscriminate helpers? Some researchers say these findings demonstrate that children, as long as they feel comfortable with the other person, will help indiscriminately (Warneken and Tomasello 2009). According to Felix Warneken and Michael Tomasello, selective prosociality does not emerge until later in life, typically around the age of three (Warneken and Tomasello 2013a). Proponents of this view suggest that from an evolutionary perspective, since children are mainly surrounded by mother and kin, there is no pressure to be selectively prosocial and it is only when the child's social circle expands to include other children that they learn who they should help and who they should avoid (Warneken and Tomasello 2009). However, as discussed in an earlier chapter, in the evolution of hominins, alloparents, such as fathers, siblings, aunts, uncles and grandmothers, were instrumental in helping with care. Recall the evidence detailed in Chapter 3 indicating that many members of the family engage in helping with childcare and that humans can be described as cooperative breeders (Hrdy 2009). From an evolutionary perspective, having a preparedness to distinguish which 'stranger' was helpful and which one was not makes better sense (Hrdy 2005).

Furthermore, other researchers have proposed that since the prosocial behaviour is directed towards adults rather than children, the child is indicating their helpfulness to kin and significant others (Wynn 2009). Selective helping may reflect an infant's motivation to be with or to signal one's helpfulness to members of their family. From an evolutionary perspective, it would make more sense to suggest that indiscriminate prosociality, even within the first year of life, is highly unlikely. Although there is no research of which I am aware of that deals with children at such a young age, there has been evidence of selective helping with children under 2 years of age.

In the first experiment, twenty-four 21-month-old infants were introduced to a pair of actors who offered, but failed to hand over, a toy to the infant. The reasons why the two actors failed to give the toy to the child were different. One of the actors failed to give the toy because she was unwilling to do so, while the second actor was unable to do so. The researchers wanted to ascertain whether the reason why the actor failed to help played a part in the reasoning process of the child. After the initial phase of the experiment, the children were directed to a new toy in the room. Both actors reached for the toy at the same time, indicating that they both wanted the toy. The toy was out of reach for both actors, suggesting that they required help from the child. The researchers reasoned that a preference by the child to retrieve the toy for one of the actors would signify that the child's choice to help one over the other was based on the earlier intentions of the actor and not the outcome (Dunfield and Kuhlmeier 2010).

Out of the twenty-four infants participating in the study, sixteen helped one of the two actors, while eight of the infants kept the toy for themselves. Among those who chose to help, three-quarters gave the toy to the unable actor rather than to the unwilling actor. This illustrates that children tend to selectively help those who indicated an intention to help them in previous encounters. In other words, for young children, it is the thought that counts and not necessarily the outcome.

In the second experiment, the researchers made a slight modification by replacing the unwilling actor with an actor who successfully returned the toy and handed it to the child in the initial part of the experiment. This experiment included twenty 21-month-old infants and, similar to the first exercise, sixteen of the infants offered the actors the toy. Importantly, there was an equal distribution of the toys between the two actors. From this, one can conclude that the infants did not necessarily base their reasoning for helpful behaviour on a successful outcome but were more concerned with the intention of the adult.

In a third experiment, the researchers wanted to see whether infants would choose to help the individual who overtly showed a willingness to provide a toy over an individual who doesn't make their intention very clear. Once again, sixteen of the infants offered the toy to the actor. Among those children who offered the toy, 75 per cent of the children preferred to give the toy to the actor who had previously demonstrated a willingness to hand over the toy more often than the actor who returned the toy but whose intention was not made explicit. This suggests that the majority of the children chose to help although their helping behaviour is even selective when 'neither of their potential recipients has previously displayed unwillingness to provide a desired toy' (Dunfield and Kuhlmeier 2010, p. 526).

These three experiments suggest that while infants have the potential for helping behaviour, when they choose to do so, they take into consideration the previous interaction with others. Furthermore, in assessing the behaviour of others, the children do not necessarily base it on the outcome but it appears that their choice to help is influenced by the intentions of other people. These experiments also suggest that infants do not necessarily refrain from helping others who previously failed to help them, but they are more concerned with the other person's intention or willingness to help. As the authors of this study note, a willingness to provide help (even if unable to) may characterize them as beneficial cooperators.

In another study, researchers used fifty-four children ranging from thirty-five to thirty-seven months to see whether, at such an early age, children would take into consideration the helpful or harmful intentions of others when determining whether the person is worthy of prosocial acts (Vaish et al. 2010). This is an important study because the children making the judgements were not themselves affected by the harmful or helpful behaviour. In other words, the children were not the victims but the witnesses to the incident. By using third-party interactions in this way, researchers can rule out emotional responses to being helped or harmed.

In the first study, the children witnessed the actor harming, helping or behaving neutrally towards a recipient. The authors predicted that the children would assess the actors' behaviour towards the recipient, and this would in turn affect the children's helping behaviour towards the actor. In the harm condition, the recipient placed her possession on the table and the actor said that she was going to damage it and then proceeded to do so. The recipient watched sadly but did not say anything to the actor. In the help condition, the recipient dropped an article and the actor helped to retrieve it. In the baseline condition, the actor simply made neutral comments about the recipient's article.

After the familiarization phase, the children watched the actor and another person play a game individually although both actors were playing at the same time. At one point, both actors reached to get a ball that was slightly beyond the grasp of each. If the child, who was watching, did not respond within fifteen seconds, they were prompted to pick up the ball and give it to whomever they chose. After this helping test, the children were handed a second ball to give it to whomever they chose. Results indicated that in the baseline condition, twelve of eighteen children helped the actor. In the helping condition, eleven of the eighteen children chose to help the actor, essentially producing the same results as the baseline. In the harm condition, significantly fewer children helped the actor, as only four out of the eighteen did so.

This study demonstrates that as early as three years of age, the majority of young children consider the harmful behaviour of others towards third parties, and this negatively affects their decision to offer help. Since the baseline and the helping conditions produced similar results, this suggests that the helping behaviour did not impact the child's behaviour. Children seem not to be impacted by the prosocial acts of others; rather, they refrain from helping only when they judge the other person to be a person undeserving of helpful behaviour. Additionally, when the children were given a second ball, even though the actor had displayed harmful behaviour, the majority of children still passed the ball to the actor, indicating that even a harmful actor can be helped if that actor is the only one in need.

It has been argued that children naturally help others, and they do so at a very young age, but they do so, discriminately. Now, I want to consider whether there is any evidence to suggest that children show signs of helping behaviour when there are no other people to encourage them to do so or even when there is no person present to thank or praise them. For instance, consider walking along the sidewalk and you find that strong winds have knocked the local cafe sign over and it is lying on the ground. For many, but not all mind you, the natural reaction is to stop, set the sign back up and carry on with your day. There is no one present to see your courteous gesture, and it is unlikely that you would update your Facebook status so everyone could see how considerate you are. You commit a neighbourly act without strengthening your reputation, receiving praise or feeling compelled to do so from a threat of punishment. It turns out, experiments with toddlers suggest they behave in the same way.

Researchers sought to determine whether children would help anonymously (Hepach et al. 2017). The question was whether children were concerned with helping others or were they more interested in receiving some type of credit or recognition for attempting to help. In the first study, there was a low rate of responses, but that may have been due to the novelty of the apparatus that was being used. Apparently, the apparatus was a device that the children were not used to using so it may have caused problems for the children. In the second study, the researchers chose a more natural situation for the children and used objects falling to the ground, as has been used in a number of other studies involving children. The researchers designed the room so that while the child was doing something for the researcher, a can would fall to the ground. The target condition was one where the adult had left the room so that she did not witness the can falling and therefore did not even realize that she was in need of help. The first control condition was a situation where the adult was present in the room but was turning away, while the second control was when the adult was absent but the situation did not call for assistance. That is, there was no mishap with the can.

The majority of children, without any adults in the room, picked up the can that fell to the ground and returned it to the table. By eighteen months of age, children clearly helped others in anonymous situations. This suggests that concerns about receiving a reward or recognition for helping behaviour are 'neither foundational nor facilitative of toddler's pro-social behaviour' (Hepach et al. 2017, p. 144). At a very young age, children are concerned with the well-being of others. It is this concern for others that is natural for humans and this is instrumental for 'thin cooperation'.

Helping and intrinsic rewards

It was demonstrated in the previous studies referred to in this chapter that children, across a number of different situations, assist others without the expectation of rewards or the threat of punishment. For instance, one study found that it was not necessary to offer material rewards to elicit help from children, nor did it appear that rewards are necessary to increase a child's tendency to help (Warneken and Tomasello 2007). Researchers have suggested that the reason children behave prosocially is, in part, driven by intrinsic motivation to help others (Hepach 2012). Some studies have suggested that the promise of rewards may not necessarily be a positive motivator, and rather than enhance self-motivation, it could have a detrimental effect (Ryan and Deci 1985). This has been called the over-justification effect (Deci 1971). The question to answer now is whether external rewards have a negative effect on intrinsic motivation. Admittedly, the answer to this question has not been settled (Cameron and Pierce 1994); however, a meta-analysis of 128 experiments strongly suggests that underlying one's intrinsic motivation to behave prosocially is the need for self-determination. The authors of the analysis conclude that circumstances that leave an individual without any sense of control will undermine intrinsic motivation, whereas situations that increase perceived self-determination will strengthen intrinsic motivation (Deci et al. 2001).

The meta-analysis to be considered presently examined whether extrinsic rewards can undermine intrinsic motivation with respect to educational policy. However, it is argued here that the results of the study support the position held in this book, namely, that an increase in cooperation depends largely on taking seriously the autonomous individual's right to lead a life of his or her choosing. This is in contrast to the BGP claim that people are motivated by threats of severe punishment for misbehaviour or the promise of rewards in a future life.

The authors of the meta-analysis distinguish between two aspects of rewards. There is an informational aspect to rewards, as it conveys self-determined competence and this feature encourages intrinsic motivation. In contrast, there is also a controlling factor in rewards that lowers perceived self-determination and therefore undermines intrinsic motivation. The authors suggest that the interpersonal context is important because if an individual feels that they are being pressured to think or behave in a particular way, then they would likely experience this as a controlling environment. In contrast, positive feedback would enhance intrinsic motivation. The authors of the meta-analysis concluded that verbal rewards, such as positive feedback, had a positive effect on intrinsic motivation, whereas material rewards had a negative effect if the interpersonal context was within a controlling environment (Deci et al. 2001). This study suggests that to enhance intrinsic motivation, a high value must be placed on the individual's freedom to choose. The evidence suggests that choice and self-determination are important for intrinsic motivation.

In another study involving children twenty months of age, researchers found that offering extrinsic rewards actually reduced helping behaviour (Warneken and Tomasello 2008). In the treatment phase, the researchers assigned children to one of three distinct conditions that differed in how an adult would respond to the helping behaviour of the child. The three conditions were material reward, praise and neutral or no reward condition. In this study, thirty-six children were in a room while the experimenter was either writing a letter or cleaning her desk. Since children are motivated to either help or engage others, the experimenters had to provide the children with distractor toys. This would force the children to give up an enjoyable activity in order to help. The experimenter would drop something, and when the children in the reward condition picked up the object they were rewarded with a toy. Those who helped in the praise condition were thanked and those in the neutral condition received no response from the adult. The authors wanted to find out whether a reward or lack of reward had any impact on the behaviour of the children.

The researchers found that children in the praise or no reward condition continued to help the experimenter almost 75 per cent of the time. In contrast, those who were given a material reward in the treatment phase helped less than the other children. These findings suggest that children have a strong tendency to help others, and that material rewards tend to undermine this motivation. From this, it can be argued that helping is rewarding in itself, but the satisfaction derived from the prosocial behaviour will be absent once the behaviour is accompanied with an external reward. Some researchers reasonably suggest that intrinsic motivation is actually dependent on whether the person has the free choice to act or not (Deci and Ryan 1985). There are a number of studies suggesting that choice has a positive effect on whether a person acts prosocially or not (Patall et al. 2008). In order to be able to understand one's freedom of choice, the children must have an appreciation of their own agency. In the study to follow, researchers specifically addressed the role of free choice in the prosocial motivation of young children.

Developmental psychologist Diotima Rapp and colleagues pointed to studies demonstrating that by the age of three, children have a sense of freedom of choice when it comes to personal matters (Nucci and Weber 1995). They also acknowledged that having a choice is likely to increase the future prosocial behaviour of children. The authors wanted to discern whether children will tend to display increased immediate prosocial motivation if they have the freedom to choose their action over being told to act prosocially (Rapp et al. 2017). The authors hypothesized that the children more motivated to help others would be those children who were provided with a choice to act.

The study included forty-eight 3-year-olds and forty-eight 5-year-olds, who were divided into two conditions. All children were in a room with a pile of paper on the floor and beside the paper was a cardboard box with a spade for cleaning up. All the children were told explicitly that they were either in a choice group or a no-choice group. The choice group could choose for themselves whether they wanted to help clean up, while the no-choice group were obligated to help, as they were told to do so by the authority figure. Results indicated that for the five-year-olds, having a choice had a positive effect on prosocial motivation (Rapp et al. 2017). In contrast, the freedom to choose had no effect on the three-year-olds. This study suggests that the freedom to choose has a positive motivating effect on five-year-olds to help others. In another study with preschoolers, it was found that freely choosing a prosocial action led to subsequent sharing behaviour (Chernyak and Kushnir 2014). Sharing is illuminating as the sharer more so than a helper incurs a cost, while the recipient benefits. I will now turn to research on the development of prosocial behaviour in children that centres on sharing behaviour.

Sharing

The above-mentioned studies on the helping behaviour of chimpanzees indicated that they simply do not help unfamiliar conspecifics. To investigate whether chimpanzees show signs of sharing with a familiar individual, researchers designed a sharing exercise where chimpanzees could choose to deliver food only to themselves or to themselves and a familiar individual (Silk et al. 2005). This sharing task was at no cost to the chimpanzee. The chimpanzee was positioned in front of an apparatus, where it had a choice to either pull a handle on one side of the device resulting in an equal amount of food for themselves and the other (1/1 option) or choose the other handle where they would receive the same amount of food, whereas the other chimpanzee would receive nothing (1/0 option). The researchers hypothesized that if there was no significant difference in the choices made by the chimpanzees, then it would suggest that they were indifferent to the welfare of others. In the eighteen chimpanzees tested, not one of them was more likely to choose the prosocial option regardless of whether another chimpanzee was present or not. The factor that led to the likelihood of choosing the prosocial option was where the chimpanzee happened to be seated. That is, the prosocial option rested with how close the handle of the device was to the chimpanzee.

Following the same paradigm used by Silk and colleagues, other researchers designed a similar apparatus to see whether children would choose to deliver the food only to themselves or share with another child (Brownell et al. 2009). The target age group for this study was 18- and 25-month-old infants. Unlike the study with the chimpanzees, the researchers in this experiment added a second component whereby the child was prompted by communicative cues. In both the original and the communicative condition, the eighteen-month-old infants chose the prosocial option at chance level. When the 25-month-old children were clearly informed that the other child needed a snack, they chose the prosocial option at above chance level. However, this was not the case if the adult said nothing. This suggests that sharing behaviour develops at around two years of age and appears that the child needs to be informed that the other is in need.

Collaboration and sharing

In an earlier chapter, I argued that as a result of evolutionary pressures, and well after the human–chimpanzee split, the line of hominins leading to *Homo sapiens* was forced to work together in order to obtain food. This led to selective pressures to share the fruits of their labour in collaborative efforts. It was pointed out that while other primates forage in small groups, only humans work together jointly to obtain food. The evidence from nonhuman primates suggested that they engaged in cooperative hunting or foraging that may be referred to as cooperation in the 'I mode', while human collaborative effort can be regarded as cooperation in the 'we-mode' (Tuomela 2002). Additionally, it appears that in all human societies, most food is obtained through collaborative efforts (Hill 2002). The question to be answered now is whether young children are more inclined to share resources if those resources were obtained from a collaborative effort or from another source that was not the result of a cooperative enterprise. If children make their choice based on the efforts of others, this would stress the importance of demonstrating to others that one is a good potential partner and fully committed to equitable sharing. From an evolutionary perspective, those cooperators who tended to share would be the type of partner that others would seek. Alternatively, if it could be shown that nonhuman primates did not adjust their behaviour in a similar fashion, it would provide support for the importance of collaborative foraging in hominin evolution. I will now turn to a study that sought to determine whether children and chimpanzees based their behaviour on the way the resources were obtained (Hamann et al. 2011).

Working with children two and three years of age, Katharina Hamann and colleagues found that when two children worked collaboratively, they shared 'much more' than when working in parallel or acquiring resources in a windfall (Hamann et al. 2011, p. 329). This suggests that as early as two years of age, children consider whether the other person did their fair share of the work and whether this affects distribution of resources. Children at this young age are sensitive to issues of distributive justice and fairness. This study implies that children do not necessarily have a fairness instinct but whether they construct a concept of fairness through working with a peer. So when children take part in a collaborative endeavour, they have the cognitive capacity to reason that the resources should be equally divided. In studies with chimpanzees, the results are much different.

Since chimpanzees do not collaborate on foraging as humans do, the researchers surmised that the chimpanzees' inclination to share would not be affected by whether the resources were obtained in a windfall or from working collaboratively. The results confirmed their assumptions as there was no difference whether the chimpanzee received the resources from a collaborative effort or obtained them by other means. Additionally, unlike findings with children, in most cases the chimpanzee kept the reward for itself. Only in 4 per cent of the cases did the lucky chimpanzee give up 'the food voluntarily by tolerating the unlucky partner taking it' (Hamann et al. 2011, p. 330). The researchers followed up with two more experiments with chimpanzees, and the results were similar. These experiments involving children and chimpanzees show clearly that the children share the food if they worked together in getting it, whereas chimpanzees do not.

The previous study suggested that children took into consideration the effort of their collaborating partner, and this develops around the second year of life. Another study involving slightly older children, between four-and-a-half and six years of age, was designed to see whether they would selectively share with friends over a familiar nonfriend or a stranger in a resource allocation game (Moore 2009). In total, sixty-six children were asked to draw a picture of a friend in their class with whom they frequently played, as well as draw a picture of a nonfriend in the class with whom they did not enjoy playing. The children were also shown a picture of a same-sexed child who they had never met. There were two different trial types in this experiment. In the prosocial trial, the child had the option of choosing one sticker immediately for themselves or wait and receive a sticker at a later time, and if they chose to wait, the other recipient would also receive a sticker. In this trial, there was no material cost to the child as he/she received the sticker in either case, but there was a delay cost. In the sharing trial, the children were confronted with the decision to either take two stickers for themselves immediately and the recipient would receive nothing or wait and receive one sticker later, but by waiting, the other recipient would also receive one sticker. This is different from the previous trial as in this case, there is both a material cost and a delay cost in choosing to be fair.

Results of this experiment revealed that children were more likely to choose the 1/1 option in the prosocial trials when the recipient was the friend or stranger than when the recipient was the nonfriend. Interestingly, there was no difference in allocating resources with the friend and stranger. In the sharing trial, the children were more likely to choose the equitable option when the recipient was the friend, but when there was a material cost, children treated the nonfriend and stranger alike. This study shows that the decision to share is, to a certain degree, influenced by the characteristics of the recipient, and depending on this information, children are willing to share, even at a cost to themselves.

Furthermore, in those situations where there are low personal costs, young children are willing to make the first cooperative move with a stranger. One interpretation of this may be that even at a young age, children assess the individual recipient and consider 'whether the potential payoff is worth the investment' (Moore 2009, p. 947). This could also mean that children are willing to share equally with friend or stranger, but if they have a reason not to share with a particular person, such as a classmate they do not like, then they refrain from sharing. What is very clear here is that it is unlikely that direction from a parent or authority figure is required for the prosocial behaviour found in these studies. As Moore points out, it is inconceivable to think that parents coach their children to be generous to friends, and as long as there is no cost to the child, they should also share with strangers.

Earlier in the chapter, I pointed to studies with twenty-month-old infants, where it was found that children are less likely to engage in helping behaviour if they had been recipients of a material reward in a previous treatment phase of the experiment, compared to those children who received praise or nothing at all (Warneken and Tomasello 2008). Other studies found 24-month-old children demonstrated helpful behaviour regardless of whether a parent was or was not in the room (Warneken and Tomasello 2013b). From these findings, it was suggested that helping behaviour in young children is intrinsically motivated, rather than the result of socialization by means of reward or punishment. A fairly recent study supports a similar conclusion with respect to sharing behaviour (Ulber et al. 2016).

In this study involving forty-eight children, ranging from thirty-six to forty-eight months in age, the children were paired with a puppet and were shown how to retrieve marbles while working together. The device used for the collaborative work was rigged so the marbles were allocated disproportionally, with the child receiving three marbles and the puppet only one. From the child's perspective, this was an accidental distribution. Based on the findings in other studies, the researchers predicted that the majority of children would share their marbles with the puppet in order to make the outcome equal and fair (Hamann et al. 2011).

The children were randomly assigned to a group receiving a material reward or a group receiving verbal praise or a third group where no response was given. In the first phase of the experiment, the children participated in the collaborative work and were treated in one of the three ways depending on the group assigned. The authors found that the children who were either praised or were recipients of a neutral response continued to equalize the disproportionate allotment of marbles, while those who received material rewards were less likely to share (Ulber et al. 2016).

In the first study, the children received the rewards with the partner who they were collaborating with, so in a second study, a neutral partner rather than the direct play partner was responsible for providing the child with the rewards or praise. The results from the first and second studies were essentially the same, in that children were less likely to engage in costly sharing if previously they were the beneficiary of a reward for the same behaviour. Additionally, the results of both experiments suggest that children reason about fairness and their behaviour to equalize the situation is intrinsically motivated. From these findings, it can be said that young children, at a cost to themselves, willingly equalize disproportionate resource allocations. Moreover, any authority or reinforcement has a negative effect on their sharing motivation. This contrasts with the BGP hypothesis that prosocial behaviour requires reinforcement or enforcement. In combination with the results from similar experiments addressing helping behaviour, these experiments suggest that the majority of children naturally want to engage, help and share with others.

Conforming

In natural habitats, a number of different species form definitive groups that display dissimilar behaviour. Studies indicate that in many species, conspecifics in a group learn from each other (Whiten et al. 2005). However, humans, unlike other species, show signs of what researchers call strong conformity (Haun and Tomasello 2011). Strong conformity entails that not only do humans conform to arbitrary fashions, such as modes of dress, but they also conform to the opinion of the majority, even when the person knows better. There have been a number of experiments demonstrating that participants knowingly gave false responses to comply with the erroneous answers given by the majority (Asch 1956; Cialdini and Goldstein 2004). For Tomasello and Haun, strong conformity has been instrumental in the transmission of human culture by encouraging quick and lasting in-group uniformity (Haun and Tomasello 2011). In the following study, researchers set out to see whether there was a difference among humans, chimpanzees and orangutans in adjusting behaviour to comply with the conduct of their peers (Haun et al. 2014).

In this study, in order to retrieve a treat, participants were tasked with placing a ball in a hole in a box that was divided into three sections. The treat for the children was chocolate, while the chimpanzees and orangutans received peanuts. During the initial phase, participants discovered for themselves that only one of the sections of the box led to a reward. After the primary phase, the participants were led to an area where they watched their conspecific peers interact with the box. The peers had been trained by the researchers to prefer only one section of the box, and if given the chance, the peers used the same section inclusively. After watching their peers, the participants were given the opportunity to play the game again. The researchers sought to see whether the participants changed behaviour to conform to the majority of their peers, stay with their original choice or go with a different option. The researchers found that children as young as two years of age tended to adapt their behaviour in order to comply with their peers. Chimpanzees and orangutans, on the other hand, were not influenced by the behaviour of their peers. While the nonhuman primates almost exclusively stayed with their own way of doing things, children conformed more than half of the time (Haun et al. 2014).

The studies presented in this chapter suggest that the majority of children naturally help and share with others, so it seems that the default behaviour for humans is to cooperate rather than to be utterly selfish. These studies run counter to the BGP claim that greater cooperation is the result of strong in-group uniformity, facilitated by a belief in punishing gods. The majority of children are inclined to conform with their peers and do not need to be threatened to do so. The evidence presented in earlier chapters, as well as the results from the experiments in this chapter, strongly support the position that humans evolved for 'thin cooperation'. Although learning and strong conformity have been beneficial for humans for in-group uniformity, one must be cautious in blessing one group or category such as race, nationality and religion with exclusive importance. Problems arise when groups identify themselves as superior to other groups. As Amartya Sen points out, much violence in the contemporary world is due to categorizing people along religious or cultural lines, and if we seriously want to achieve peace, we need to recognize 'the plurality of our affiliations and in the use of reasoning as common inhabitants of a wide world rather than making us into inmates rigidly incarcerated in little containers' (Sen 2009, p. xv). In other words, in order to achieve 'thick cooperation', humans must overcome their natural inclination to act prosocially only within the in-group.

Moral realism

In previous chapters, I summarized and critiqued the central claims made by the BGP. To briefly recapitulate, the BGP regard the rise of large-scale human societies as an evolutionary puzzle since, according to them, such cooperation cannot be explained by kinship, reciprocity or reputation. Therefore, they reason, an alternative explanation is required. Central to this alternative explanation is the development of cultural evolutionary processes that are primarily driven by competition between groups. Over time, this competition resulted in groups gradually amassing packages of cultural elements that enabled group solidarity and facilitated the rise of large-scale cooperation. For these authors, 'our ability to entertain supernatural and ritualized practices, as by-products of our evolved cognitive capacities, provides the foundation for the rise of prosocial religions and of complex, cooperative societies' (Slingerland et al. 2013, p. 338).

The BGP suggest a belief in Big Gods allowed the moralization of human action in particular ways, as well as to 'incentivize certain behaviours using supernatural rewards and punishment' (Slingerland et al. 2013, p. 336). They also suggest that supernatural agents manipulate our psychology in such a way that it provides a fitness advantage in competition with other groups (Slingerland et al. 2013, p. 336). For the BGP, it was the accumulation of these cultural packages that facilitated the origins of complex, large-scale societies, and those societies adopting beliefs in moralistic gods outcompeted societies that adopted other types of religious beliefs and practices. They invite us to believe that it is the connection among religion, group identity and morality that provides a 'cultural evolutionary explanation for the emergence of the moral realism that now pervades both religious and secular discourses' (Slingerland et al. 2013, p. 337). For the BGP, in order to sustain norms, successful groups postulated norms that originated with supernatural agents or at least provide them with some sort of supernatural authority.

In order to demonstrate that moral realism emerged by granting authority to Big Gods or by tracing norms to the metaphysical structure of the universe, the BGP adopt Charles Taylor's distinction in human judgement between strong and weak evaluations. The former draws their power from 'being based on one or more explicit or implicit metaphysical claims, and are therefore perceived as having objective force, rather than being a merely subjective whim' (Slingerland et al. 2013, p. 346). The BGP suggest that, in this way, people are motivated to punish any contravention of strong evaluations and condemn such violations in metaphysical terms.

To demonstrate the distinction between strong and weak evaluations of judgement, the BGP solicit us to think of a person who prefers chocolate over vanilla ice cream. They offer this as an example of weak evaluation as the person would not expect this food preference to be universal, and a person with this inclination 'is certainly not moved to condemn others for preferring chocolate' (Slingerland et al. 2013, p. 346). They go on to suggest that people in modern Western societies are not inclined to sexually abuse small children, but they note that this is an entirely different sort of preference. Abusing young children just feels wrong, and people are moved to punish anyone who violates this feeling. According to their hypothesis, if people are pressed on the matter, they frame their conviction in metaphysical terms, that is, the condemnation of child abuse is based on 'beliefs about the value of undamaged human personhood or the need to safeguard innocence' (Slingerland et al. 2013, p. 346).

The BGP are making the claim that in order for moral realism to have a sustainable objective force, the norms must be backed up by a god-like authority. This suggests that moral realism is dependent on a belief in a supernatural power. Yet researchers investigating moral development clearly show that even very young children naturally accept moral objectivism (Nucci 2001). One question to consider is whether this moral objectivism is an innate concept of moral violations 'which carries with it a commitment to moral objectivism' (Nichols 2003, p. B31). Studies by social domain theorists suggest children find moral prescriptions as 'objective' in the sense that they are moral even though they are not contingent on an authoritative figure. The studies by social domain theorists demonstrate, quite clearly, that morality for most children, across all cultures, focuses on the interpersonal issues relating to justice and human welfare (Nucci 2001). In contrast to the assertions made by the BGP, children's reasoning about moral issues is independent of religious moral doctrine. I will now discuss some of these studies to counter the BGP claims.

Psychologist Elliot Turiel has shown that there is a distinction between violations of moral principles, such as sexually abusing a child, and violations of conventional rules, such as texting during class (Turiel 1983). If there is no rule or convention prohibiting texting in the classroom, then there would be no violation of the convention. Conventions are arbitrary, while moral transgressions remain transgressions even if there are no explicit instructions prohibiting the behaviour. There are certain rules particular to religions that are regarded by the followers as essential and binding; however, these rules are considered nonmoral in the social domain model.

Morality, according to the Social Domain Theory, involves interpersonal relations concerning justice, rights and human welfare (Turiel 1983). Prescriptions related to these types of acts are not dependent on the view of an authority figure. For example, the prescriptive force of the claim that it is wrong to throw a baby across the room is objective in that the result of the act is independent of the subjective opinions of the observers. While child abuse is certainly an assault against the person, there are a number of other issues dealing with human welfare that do not necessarily deal with harm. What all these issues have in common, however, is that when making judgements about moral concerns, children frame their reasoning around their understanding of human welfare and not necessarily what the authorities say. The reasons given are not concerned with 'safeguarding innocence' but are out of concern for the welfare of the individual.

Conversely, there are a number of issues that deal with social conventions, such as many of the customs particular to religions. These are agreed-upon customs for social behaviour determined by the particular culture in which they are formed (Turiel 1983). When questioning whether conventions are right or wrong, it is done so in terms of norms and authority. The normative force of conventions holds only within the particular culture in which the rule was created (Turiel 1983). In judging moral issues, decisions are based on the harm that the particular action caused. Furthermore, the distinction between conventions and moral issues is cross-cultural (Helwig and Turiel 2017).

The evidence suggests that in order to justify conventions or what the BGP call weak evaluations, such as what one should eat, wear or what type of music to listen to, or even whom one sleeps with, an appeal to the gods or supernatural agents is commonly used to justify the rule. For instance, if a ruler wants to make a convention prohibiting vanilla ice cream because she prefers chocolate, then to justify the convention and give it prescriptive force, the ruler may frame it in metaphysical terms. If people choose not to follow the convention, they risk punishment from the authorities. Food taboos that are nonmoral conventions, yet important for many religions, are justified by appealing to the gods. Historically, people have been killed for so-called weak evaluations because the merit of the convention was justified by supernatural agents. With moral transgressions or what the BGP term strong evaluations, there is no need to appeal to a supernatural authority because we naturally find certain behaviours wrong. In the studies to follow, we will find that children, regardless of what the authority figures or God says, believe that moral transgressions are wrong and that these transgressions are not just wrong for their particular religion or culture but are held universally for other religions and cultures as well.

Psychologist Larry Nucci asked Catholic adolescents and young adults about behaviour, considered sinful, according to the doctrine of the Catholic Church (Nucci 1995). Some of the behaviours involved harm or injustice to another person, such as rape, murder or theft, and these types of offences were classified, by the author, as moral matters. Other behaviours, such as masturbation, premarital sex and the ordaining of women, breached the doctrines of the Church and were considered nonmoral, analogous to social conventions. Findings demonstrate that both high school and adolescent Catholics found moral transgressions much more serious than contravening the rules of the Church. The author of the study pointed out that the students made this decision despite the fact that breaking Church rules resulted in the same severe punishment as engaging in illegal moral acts. If the outcomes for moral transgressions and violations of Church doctrine lead to the same severe punishment, then one would think that there should be no difference in the student's judgements.

Nucci suggests that Catholics judged the moral transgressions to be more serious because conducts such as assault, rape and theft have intrinsic effects on the recipient of the act (Nucci 1995). Findings indicate that 91.6 per cent of high school and 98 per cent of university students considered that even if Church leaders chose to remove prohibitions against theft and assault, that type of behaviour would still be wrong. Conversely, only 40.8 per cent of high school and 32.7 per cent of university Catholic students thought it be wrong if the Pope or other authority figures removed such nonmoral conventions, such as the ordination of women, using contraceptives or engaging in premarital sex. This counters the BGP claim that ideas about the prescriptivity of moral actions are dependent on the authority of supernatural authority. In fact, the Catholic students in this study found certain behaviour wrong regardless of what the conventions hold.

Findings also demonstrate that the Catholic students tended to universalize only the behaviour concerning moral issues and not Church conventions. Similar to the earlier results, 91 per cent of high school students and 97 per cent of university students thought that it would still be wrong if members of another religion engaged in behaviour that was assaultive or immoral in nature even if the other institution did not have rules preventing such behaviour. In contrast, only 33.8 per cent of high school and 18 per cent of university students thought that the conventions of their Church should be universalized so that acts contravening Catholic doctrine by non-Catholics would be viewed as wrong.

In another study, the researchers had two sets of questions for members of the Amish-Mennonite and Dutch Reform Calvinist Christian community, and for Orthodox Jewish children and adolescents (Nucci and Turiel 1993). The first set consisted of a series of questions asking whether the right or wrong of a certain behaviour was contingent on the authority of God. The second type of question related to, what the authors called, the open question, which related to asking whether a command from God could make right something that most people consider morally wrong, such as assaultive behaviour. The authors suspected that even very devout children would dismiss the claim that a command from God would make something like assault right and that the children would reject the claim that God would ever make such a command in the first place. The researchers thought the second question was significant as the participants had strong beliefs in the compelling nature of a command from God. However, the researchers were only permitted to pose the open question to the Dutch Reform Calvinist children.

The study consisted of three sets of questions about four moral issues involving stealing, hitting, slandering and damaging another person's property. There were also questions dealing with seven nonmoral issues, including day of worship, working on the Sabbath, women wearing head coverings, baptism, women preaching, premarital sex between consenting adults and interfaith marriage. The first questions dealt with 'alterability' as the subjects were asked, 'suppose all of the members of the congregation and the Minister agreed to [alter/eliminate] the rule about [the act] would it be wrong or all right for them to do that? Why or why not?' (Nucci and Turiel 1993, p. 1478). The second set of questions dealt with generalizability as the children were asked, 'Suppose that in another religion they don't have a rule about [the act] would it be wrong or all right for them to [engage in the act] in that case? Why or Why not?' (Nucci and Turiel 1993, p. 1478). The third set of questions, related to the contingency of God's command, asked the question, 'suppose there was nothing in the bible about [the act], God hadn't said anything about [the act] one way or another, would it be wrong or all right for a Christian to [engage in the act]? Why or Why not?' (Nucci and Turiel 1993, p. 1478). The second portion of the study, restricted to only the Dutch Reform Calvinist participants, asked, 'suppose God had commanded (written in the bible) that Christians should steal. Would it then be right for a Christian to steal?' 'Do you think God would make a commandment saying Christians should steal?' (Nucci and Turiel 1993, p. 1478).

The questions relating to alterability revealed that children steadily stated that it would be wrong for the members of the congregation and the Minister to eliminate rules that prohibited transgressions in the moral domain. Unlike the Catholic children in the study mentioned earlier, the Amish/Mennonite children thought it was just as wrong for the congregation to alter the prohibition against work on Sundays as it was to alter the prohibition against moral issues. While the range for saying that it was wrong to alter the prohibitions against moral transgression was between 92 and 95 per cent, the range for nonmoral transgressions went from a low of 41 per cent concerning premarital sex to a high of 92 per cent for working on the Sabbath. Importantly, approximately 40 per cent of the Amish/Mennonite children justified their reason for objecting to any alteration in the rules regulating moral actions based on the fact that the behaviour was hurtful and unjust.

From the questions relating to generalizability, the authors found that behaviour involving moral transgressions was determined by over 90 per cent of the children to be wrong even for members of a different religion that did not have any rules prohibiting such a behaviour. For moral issues, the range was from a low of 89 per cent for property damage to a high of 100 per cent for stealing. In contrast, far fewer children thought it was wrong to engage in nonmoral behaviour if the other religion did not have any rules governing that type of behaviour. The responses went from a low of 16 per cent for women's head coverings to a high of 50 per cent for working on the Sabbath. Clearly, the children considered behaviour within the moral domain as universal and religious norms as relative to the particular religion. Furthermore, the majority of children justified their reasons based on the effect it would have on the welfare of others and not because God commanded it.

The third set of questions is particularly illuminating for the purpose of this chapter as the questions focused on whether children believed moral behaviour was dependent on the word of God. Very few children considered it wrong to engage in the nonmoral behaviours if God had not commanded any prohibitions against such acts. In fact, with respect to the questions concerning the day of worship, working on the Sabbath, baptism, wearing head coverings and interfaith marriage, not a single child believed it would be wrong had there not been a prescription by God against it. The children justified their reasons by suggesting that if there was no prohibition by God, then that type of behaviour was fine. In contrast, when it came to questions about stealing, slander, hitting and damaging property, 84, 84, 88 and 91 per cent, respectively, of the children said that this type of behaviour would be wrong even if it was not found to be prohibited in scripture or by a commandment from God. The reasons that the children used to justify their claims centred on the fact that this type of behaviour was hurtful to another person. This clearly demonstrates that even for the most devout, prosocial behaviour stems from something other than transcendental prohibitions or the word of God.

On the 'open question', 68 per cent of ten- and twelve-year-old and 81 per cent of the fourteen- and sixteen-year-old Dutch Reform Calvinists rejected the suggestion that God's commandment to steal would make it right to steal. They justified their decision that stealing is still wrong even should God have approved it by pointing to the effect that this type of behaviour would have on others. Additionally, nearly all of the children who participated believed that

God would not make such a command because that would be in contravention of the perfection of God. Just as the Amish Mennonite children, the Dutch Reform children conceive moral issues in terms of the welfare of others and not with what religious authorities say or the word of God.

In the second study, the researchers interviewed a group of sixty-four conservative Jewish children and a second group consisting of thirty-two orthodox Jewish children from Chicago. The methodology was the same as in the previous study, and the results essentially mirror those findings. For the alterability questions, there was very little difference between the moral and nonmoral issues. However, approximately 75 per cent of the Jewish children justified their disapproval of the elimination of a rule governing moral action by pointing to the effect such behaviour would have on the other person. For the nonmoral issues, the majority of the children justified their assessment on the fact that it was God's will.

Results from the third set of questions that focused on whether Jewish children believed that moral behaviour is dependent on the word of God are again unambiguous. Indeed, very few Jewish children considered it wrong to engage in the nonmoral behaviour if there was no commandment from God to prohibit such behaviour. For instance, questions regarding day of worship, work on the Sabbath, men's head covering and women reading from the Torah, not a single child from either the conservative group or the orthodox group saw it as wrong if there was no command from God prohibiting it. Of the conservative Jewish children, only 14 per cent considered interfaith marriage still wrong and 38 per cent of orthodox Jewish children found that premarital sex was still wrong even if the rule did not come from God. In clear contrast, 91, 94, 97 and 97 per cent, respectively, of the conservative children found slander, stealing, hitting and damaging property as still wrong. For the orthodox Jewish children, 97, 97, 100 and 100 per cent, respectively, found stealing, damaging property, slander and hitting as still wrong even if God did not prohibit it.

From these studies it is quite clear that Christian and Jewish children make decisions about moral issues independent of the word of God or what religious authorities have to say. The objective criteria for these children were how the behaviour would impact the rights of others. They were concerned with the harm caused to others and not with any metaphysical claims. For the devout children in this study, the idea of a punishing God didn't 'incentivize' their behaviour; rather, they reasoned about the questions and based their answers on a concern for the rights and welfare of others.

Let's now consider the evidentiary value of this study and compare it with the studies involving the priming literature that was discussed in the second chapter. In the studies in this chapter, we find children reasoning about both moral and nonmoral issues. The evidence obtained is derived from what the children are telling the researchers and not what the researchers believe the children are thinking. In contrast, in the priming literature offered by the BGP to support their hypothesis, the conclusions they draw are based on what the researchers think the participants may have been thinking at the time of the experiment. For instance, recall the study where participants were asked to unscramble '*dessert divine was fork the*' and presumably by rearranging the sentence to read 'the dessert was divine' the participants were primed to think of punishing, moralizing gods (Shariff and Norenzayan 2007). As suggested, many of the words could mean a number of different things to people, yet the BGP conclude that these words 'incentivized' the behaviour of the participants to act prosocially. For the BGP, the so-called religious terms not only 'incentivzed' participants to think about God, but the participants were also specifically thinking of a punishing, moralizing God. Clearly, the studies in this chapter are much more persuasive.

Conventional norms

In the above studies, it is quite clear that children reason about moral issues and make decisions independent of what God or an authority figure has to say. There have been a number of studies demonstrating that children from the age of two enforce conventional norms on other children by protesting against any norm offender (Rakoczy et al. 2008). The rules in these other studies were imposed by an adult or another authority figure. Recently, researchers sought to see whether children would treat their self-created rules differently than rules that were commanded by adults (Hardecker et al. 2017). Would children have the same understanding of the normativity of their self-created rule and would they show the same flexibility with their own rules and the rules imposed by an authority figure? The authors also wanted to compare children's enforcement of their self-created rules and rules given by adults. They also wanted to see if the children would show greater flexibility in their own created rules. A number of interesting findings came out of this study.

The first study included 115 children between the ages of four and five. There were two conditions in this study. One was called the invention condition, where children were asked by the experimenter to invent a game so that he or she could use it to play with other children. The second group of children were categorized as the discovery condition, where the experimenter told the children that someone had told him or her how to play the game and he or she would show the children the 'correct' way to play.

Although not a focus of the present study, the researchers noted that all twenty groups in the invention condition negotiated the game rules together, and the children took turns making suggestions and either agreeing to or rejecting the suggestions of others. In sixteen groups, there was at least one child who led the group by asking whether everyone was in agreement with the rule. This suggests that the children understood the task as a collective undertaking and that the rules they devised required agreement. It also indicates that within groups, there tends to be at least one person who takes charge.

Interestingly, in both conditions, the majority of children used normative language to teach the game and there was no difference between the two conditions. In other words, regardless of whether the rules were created by themselves or were imposed on them by an adult, the children used the same normative language. They also found that there was no significant difference in the likelihood to enforce rules when third parties violated the rules. This is significant as it suggests that children 'extend the normative scope of arbitrary rules they create themselves to third parties' (Hardecker et al. 2017, p. 174). To the surprise of the experimenters, children who received the rules from the adult showed greater flexibility than they had suspected. This finding is consistent with the work of the social domain theorists who contend that preschoolers are quite flexible in their consideration for changes in conventional rules but not moral norms (Turiel 1983). The authors surmise that they would have found less flexibility to rule change if it included moral issues.

In a second study with seven-year-old children, the findings were similar. Interestingly, the older children allowed fewer changes to their self-created rules. This may be due to the fact that older children take their rules more seriously and the fact that since this was a collective effort perhaps they saw this on the same level as adult rules. Nevertheless, the two studies demonstrate that normativity for children comes from two different sources and one has no greater authority than the other. The first is norms handed down from adults or another authority figure, and this normative standard applies to everyone in the group. Norms can also be created by collective negotiation and mutual agreement among the children in the group. Since it is a collective agreement, children come to understand that their agreement to the rule is a commitment that gives the rule an objective force. Finally, these findings suggest that children's understanding of normativity is not simply as rules imposed on them by adults in which they must conform. In fact, children find the rules that they have self-created by negotiation and reasoning just as binding as the rules that are handed down from authority figures.

Helping behaviour despite the rules

To end this chapter, I will move from experiments with children to that of university students. In an interesting study to be considered presently, the main purpose of the experiment was to set up a situation in order to determine whether psychologically based interrogation techniques would likely elicit true or false confessions. Nevertheless, this experiment provides additional evidence to demonstrate that people don't necessarily comply with rules because of a fear of punishment and at times are willing to break a rule in order to help another person (Russano et al. 2005). The experiment involved guilty and innocent participants being accused of intentionally breaking a rule or cheating. The researchers hoped to demonstrate that certain police tactics could increase the rate of both true and false confessions. What is important for this chapter is what occurred during the first phase of the experiment and not with the interrogation methods used in the second part.

In this experiment, 330 undergraduates were selected to participate and fourteen were excluded because they revealed a high level of suspicion and two others declared a wish to abandon the project. The participants were advised that they were taking part in an important study to assess the differences between individual and team decision-making skills. Each participant was matched with a confederate and after a short rapport-building period, the experimenter told the pair that they should only work together on designated team problems but must work individually, without collaborating with each other, on the individual problems (Russano et al. 2005, p. 483).

Half of the participants were assigned to the guilty condition, where the confederate asked for help with the individual problem, and the other half were in the innocent condition, where the confederate did not seek assistance. Interestingly, the experimenters had many subjects to use in the interrogation segment of the experiment, as the vast majority of the participants helped the confederate on the individual problem. In fact, there were only eighteen people who did not help the confederate with the individual problem. I suggest that the reason why so many students 'cheated' was simply due to the human inclination to assist others. Despite the fact that there was a rule prohibiting students from helping in the individual problem, the vast majority chose to help anyway. I suggest that the students were inclined to cheat because they assessed the situation and reasoned that helping someone outweighed the rule that prohibited it. Importantly, the people whom the participants were helping were strangers. This suggests that the majority of people have difficulty *not* helping others. Clearly, this is in direct contrast to the suggestion that people need to be threatened to act prosocially.

Granted, it is true that cheating is frowned upon in society and there have been some researchers who have persuasively argued that humans have a cheater detection device as this would have been evolutionarily adaptive (Cosmides and Tooby 1992). However, people find cheating offensive when the cheater is the sole beneficiary of the act at the expense of an innocent person. It is even more problematic from the perspective of the person who was deceived. In this experiment however, the cheater was not the beneficiary; rather, the participant's cooperative behaviour was at a cost to himself and there was nothing to be gained from the decision to help the other individual. This study, as well as all the experiments outlined in this chapter, clearly shows that humans have a natural inclination for 'thin cooperation'. In the final chapter, I will review the evidence presented in this book and provide a final explanation as to how to achieve 'thick cooperation'.

6

The road to 'Denmark'

Historian Ian Morris points out that today, across the planet, a person is less likely to be violently killed than in any other period in history (Morris 2014). This may come as a surprise to many, but statistics suggest the world is less violent and there is greater cooperation today than ever before (Pinker 2011). For instance, in 2012, one person in 4,375 was violently killed. From this, it can be deduced that about 0.7 per cent of the people alive today will die violently. In contrast, 1–2 per cent of the people who lived in the last century died at the hands of someone else, 2–5 per cent for those living in ancient empires and a staggering 10–20 per cent for people living in the Stone Age (Morris 2014, p. 333). Clearly, we are much better at getting along.

What is 'Denmark'?

In their paper 'Solutions when the Solution is the Problem', economists Lant Pritchett and Michael Woolcock suggest that most people would agree that the long-term goal of modern political institutions is to provide people with protection, clean water, education and adequate public health – in other words, to improve the quality of their lives. This enhancement is guaranteed by effective rule-based, accountable public agencies, resembling Weberian bureaucracies (Pritchett and Woolcock 2004). The authors named this type of society 'Denmark', and they used the scare quotes to differentiate the mythical 'Denmark' from the Scandinavian country of the same name. 'Denmark' was the 'common core of the structure of the workings of the public sectors in countries usually called developed' (Pritchett and Woolcock 2004, p. 192).

The struggle to create such institutions has been so long and difficult that people who are lucky enough to live in the countries with those types of institutions suffer from 'historical amnesia regarding how their societies came to that point in the first place' (Fukuyama 2011, p. 14). Fukuyama argues that not even the Danes themselves know how they got to 'Denmark'. After all, the Danes descended from the Vikings who conquered and pillaged much of Europe and were a tribal group who, like most groups in history, owed allegiance to their kin and in-group and settled problems through retributive justice, rather than a fair rule of law. Currently, the real Denmark is one of the safest and most cooperative countries in the world, where in 2009, just one person per 111,000 was murdered (Morris 2014). That represents a risk of violent death of just 0.027 per cent, and the world continues to get more and more 'Danish'. The argument to unfold in this chapter contends that the way to 'Denmark' did not develop by envisioning a perfectly just society and then advancing towards it. Rather, the historical evidence suggests that the road to 'Denmark' was paved by brave people fighting for the rights of others in order to make the world a little less unjust for more and more people. While the BGP may not necessarily suffer from 'historical amnesia', their theory makes little sense when we consider how it came to be that marginalized people gained a greater voice in their lives, a voice that is essential if we seriously want to speak about 'thick cooperation'.

As has been previously stated, the BGP suggest that, for most of human history, the driving force behind large-scale cooperative groups was a belief in and fear of Big Gods. Norenzayan believes that an 'elaborate but fragile set of norms and institutions' has emerged providing enforcement mechanisms that encourage trust among strangers (Norenzayan 2013, p. 170). Without these enforcement mechanisms, their theory suggests that large-scale civilization would not be possible. According to the BGP, the cultural bridge between small-scale human societies and secular societies in parts of the modern world was a belief in and fear of Big Gods (Norenzayan 2013; Norenzayan et al. 2016a). Accordingly, secular cooperative societies have passed a threshold, no longer needing belief in Big Gods to sustain such large-scale cooperation, as 'Denmark' is an outgrowth of prosocial religions such that 'secular societies climbed the ladder of religion, and then kicked it away' (Norenzayan 2013, p. 132).

God and government

According to the BGP, 'God and government are both culturally and psychologically interchangeable' (Norenzayan et al. 2016a). The BGP suggest that religious prosociality was once an effective mechanism to encourage exchange among strangers, as well as organize them for cooperative endeavours. Since strong secular institutions also make public trust possible, the selective forces for a belief-crutch began to decline. They offer as evidence the priming experiments where reminders of secular moral authorities, such as police and courts, led participants to behave more prosocially in economic games just as participants were more likely to be more generous when primed with Big God terms (Shariff and Norenzayan 2007). However, as argued in Chapter 2, being primed with science or superhero words also led to greater prosociality, so we should not place a great deal of weight on those types of experiments. Notwithstanding the lack of evidence from the priming literature, a clear difference between secular inclusive institutions, encouraging human rights, and the free exchange of ideas is fundamentally different from the theory proposed by the BGP, and this becomes especially clear once the BGP integrate 'sacred values' into their framework.

The BGP propose a number of elements that led to a group's successful expansion, one of which was 'in-group ("ethnic") markers to spark tribal psychology, exclude the less committed, and made religious boundaries' (Norenzayan et al. 2016a, p. 6). They go on to stress the importance of 'seeing a divine origin in certain beliefs and practices transforming them into "sacred values" that are *non-negotiable*' (Norenzayan et al. 2016a, *my italics*). They do acknowledge that the boundary markers set by the prosocial religions have a dark side, as they may embody political and economic inequality within the cooperative group. Furthermore, these boundaries may also contribute to exploitation by the more powerful, and it could be toxic for those who lie outside the imagined moral boundaries of the in-group. They also freely admit that within this framework 'intragroup cooperation can readily feed into intergroup antagonism' (Norenzayan et al. 2016a, p. 13).

Despite recognizing the obvious problems associated with their argument, the BGP still state, 'our approach suggests that cultural evolution anchors certain kinds of norms or beliefs – those favouring success in intergroup competition – to a kind of metaphysical bedrock such as the desires of a widely accepted and omnipotent deity' (Norenzayan et al. 2016a, p. 14). Their reasoning for the necessity of metaphysically grounded, group beneficial norms is that without such a force, cooperation would break down due to defection. Moreover, they note that large-scale cooperative societies, especially multi-ethnic societies, could be at risk of inter-group competition, leading to the breakup of the larger society. The fear is that subgroups may be motivated to counter social norms that favour the subgroup at the expense of the larger group. These researchers suggest, 'if norms are grounded metaphysically, however, self interested individuals or subgroups pushing to alter norms face a substantial obstacle' (Norenzayan et al. 2016a, p. 14). They are quite correct in their observation that metaphysically grounded norms are difficult for subgroups to challenge, but the fact that individuals or subgroups are prohibited from making claims about social norms is, I contend, the very reason why we should discourage nonnegotiable norms that are grounded metaphysically.

The BGP theory rests with a neo-Hobbesian view of human nature. They assume that selfishness and the irresistible urge to freeride are the default behaviours of humans; therefore, there is a need for Big Gods to ensure norm compliance. Recall that Thomas Hobbes believed human beings were such a nasty lot that there had to be some type of mechanism put in place to curtail selfish behaviour, and he concentrated on determining what would be the best means to accomplish such a goal. For Hobbes, as well as other social contract theorists, the most effective way of succeeding was to identify the most appropriate institutional arrangements for society. Amartya Sen calls this approach transcendental institutionalism, and he suggests that it has two specific features (Sen 2009).

Transcendental institutionalism

Sen argues that the transcendental approach starts with identifying what a perfect and just society should be like, rather than taking a comparative approach to justice and injustice. The transcendental approach is aimed at identifying 'sacred values' or identifying the nature of a just society rather than inquiring whether some alternative arrangements for society can be made to make life for people a little less unjust. Second, in order to identify the most appropriate institutions to strive for, there has to be some consideration as to how people behave and interact with each other. Inevitably, the transcendental approach leads people to make assumptions about human behaviour in order that it fits with the conception of the preferred institution.

According to Sen, both these features depict a contractarian manner of thinking. Conceiving society as chaotic, unless some ideal alternative is put in place, is the main component of the 'social contract' arrangement of society. Sen counters that it would be more effective to replace the idealistic approach of striving for a mythical transcendent just society with a comparative approach by way of focusing on societies that already exist and advocating for making life better for people. For Sen, a realization-focused comparison approach aims at the removal of injustices in the world (Sen 2009). The difference between these two approaches is extensive and can affect our understanding of how injustices in the world have been removed, as well as identifying the unfair treatment in society today.

Admittedly, the BGP are not identifying perfect institutions, but their theory has the same two features of transcendental institutionalism identified by Sen. For instance, the BGP suggest that norms have to be stabilized, in that we must transform beliefs and practices that are 'non-negotiable' and bestow them with divine authority (Norenzayan et al. 2016a, p. 6). This approach fails to engage in questions about advancing the welfare of people because it does not allow comparing alternative proposals outside of the imagined nonnegotiable beliefs. Second, they take for granted that humans are 'overly selfish, egoistic, greedy, lazy, offensive, domineering, or otherwise in violation of social norms and taboos' (Johnson 2016, p. 144). I will briefly recapitulate why they are wrong in their portrayal of human nature and then provide examples to illustrate why nonnegotiable 'sacred values' are problematic and that the recognition of such 'sacred values' does not provide opportunities for 'thick cooperation' but, at times, has been a major stumbling block against it.

Shared intentionality

In Chapter 3 I argued that all primates live in groups. Because of the selective pressures on emerging humans associated with attempting to collaborate and communicate with others, humans evolved thinking processes that can be termed a 'shared intentionality' (Tomasello 2014). That is, there is a clear distinction between humans and our simian cousins in this regard. It is humans and not apes who communicate informatively, teach each other helpfully and engage in cooperative childcare, and it is only humans who establish norms and institutions. As I also argued, humans are fundamentally cooperative, while apes are mainly exploitative. One of the main points in Chapter 3 is that fitting in with one's group and getting along with others had fitness advantages in the type of environment that humans found themselves in for most of our species' evolutionary history. In other words, humans are wired for 'thin cooperation'.

Helping and sharing

In the last chapter, it was argued that children have years of joint engagement between them and adults, and this led Vasudevi Reddy to emphasize the need to take seriously the emotional relation and psychological awareness as a starting point when it comes to understanding human minds (Reddy 2008). Joint action and shared experience happen prior to the child having the ability to communicate any private experiences. With this foundation, the child initiates joint action and does so beyond the adult–child relationship. This way of viewing the mind is consistent with Tomasello and Searle's idea of 'shared intentionality' and is supported by the evidence from the studies with children summarized in the last chapter.

For instance, it seems quite clear that the majority of children are not the Hobbesian creatures suggested by the BGP. Studies presented in the last chapter suggest that at a very young age children naturally help and share with others. This helping behaviour is proactive in that the behaviour is not necessarily in response to a cue from other persons, or the result of a threat of punishment. It is also unlikely that the helping behaviour is the result of an empathetic response because in some studies, the children naturally assisted the other person without eye contact or even without any person in the room (Warneken 2013a). Although children seem innately to help others, in some studies we found that children were selective of whom to help or share with based on their understanding of the motivation of the intended recipient (Vaish et al. 2010). For instance, in one study, children were not necessarily impacted by the prosociality of others but refrained from helping those who they understood as underserving of helpful behaviour. Although in these studies children are basing their decisions on the actions of the other people and judging whether they should help or not, it nevertheless suggests that people could be easily influenced about what other's think of the particular person or group of people. Recall that there are studies suggesting that children conform with the group, so it is not inconceivable to conclude that a person's natural instinct to help others may be cut short due to the particular ideology of the in-group. This suggests that while 'thin cooperation' comes naturally, 'thick cooperation' is much more difficult to attain.

Finally, the BGP claim that in order for groups to be successful in outcompeting other groups, moral claims need to be fortified by a god-like authority. For some of these authors, the more violent the threat from the gods, the better is the effectiveness for norm compliance (Norenzayan 2013; Johnson 2016). Yet in Chapter 2, we found that the evidence to support this claim is questionable, to say the least. Furthermore, in the last chapter, we found that the vast majority of children described stealing and violent crime as wrong, and the reason that they found it wrong had nothing to do with the word of God or a threat of punishment. In fact, the majority of very religious children found assault and theft wrong despite what God or any authority figure had to say about it. In contrast, nonmoral issues, such as food taboos or the day or time designated for worship, did not seem to be such a big deal, unless it came as a command from God. Rather than a need for nonnegotiable norms regarding violations of bodily integrity, it seems quite clear that the majority of children find those acts wrong regardless of whether they are packaged in metaphysical terms. To wrap up the argument in this chapter, I will now outline how a small group of activists pushed to alter norms, overcoming the substantial obstacles within their group, and convincing their group and the rest of the world that slavery was unjust. This is the path to be taken in order to achieve 'thick cooperation'.

Burying the chains

In his brilliant book *Bury the Chains*, journalist and academic Adam Hochschild traces the first grassroots human rights movement that began on 22 May 1787 in Britain. The citizen's movement was cleverly organized by twelve driven men who fought for the freedom of slaves around the world. Hochschild makes it clear that all these men were deeply religious but that their new faith was in the autonomy of all human beings. He said that the campaign was formed with the belief 'that the way to stir men and women to action is not by biblical argument, but through the vivid, unforgettable description of acts of great injustice done to their fellow human beings' (Hochschild 2005, p. 366). He goes on to suggest that the abolitionists placed their hope not in sacred texts, but in human empathy. By briefly looking at the grassroots movement described by Hochschild, we find that the abolitionists strategically organized their campaign and, through argumentation and reason, persuaded people that slavery was unjust. While the abolitionists may have been motivated by empathy, the crusade to end slavery was justified by skilfully articulating their cause.

In the same year that the twelve men met at a printing shop in London, nearly everyone in the country thought slavery was normal, and if anyone thought otherwise they would have been seen as delirious. Most people in eighteenth-century Britain time intuitively 'knew' that people with darker skin were inferior. So, it is quite remarkable that the antislavery campaign managed to convince so many people to fight for someone else's rights. Not only were they fighting for the rights of others, but they were also concerned for people who were from a different country and were visibly different from them. Yet in a relatively short period of time, public opinion in many countries changed.

Hochschild describes Codrington Plantation in Barbados as a typical plantation where the slaves were forced to work. The diaries from the people connected to Codrington say that the slaves worked six days a week, starting their day between five and seven in the morning and ending at seven in the evening. Children from about seven years of age were expected to work and they were given a child-sized hoe in order to help with the weeding and planting. According to one of the local ministers, the Codrington slaves spent their only day off in 'vile and vicious practices ... either in trafficking with one another or in dancing, or in other actions improper to mention' (Hochschild 2005, p. 64). The managers at Codrington would pay a gallon of rum to anyone who returned a runaway slave and, as a preventative measure to discourage desertion, they used a red-hot iron to burn on the chest of the slaves the word Society. Society was a short form for *The Society for the Propagation of the Gospel in Foreign Parts*, whose governing board included the Archbishop of Canterbury, as the Codrington plantation was owned not by individuals but by the Church of England. This suggests that slavery was normal for all people, including the most influential and powerful in society.

While the Church of England was a keen supporter of the slave trade, one small religious denomination officially expressed opposition. The Quakers believed that every human being, regardless of race or class, was a creature of God and should be treated equally. Unfortunately for the Quakers, although gifted with great organizational skills and tremendous drive, they lacked credibility in the eyes of other Britons. What they needed was an influential Anglican to be on their side in order to give them some authority. In 1784, they got what they needed when the Vice Chancellor of Cambridge, Anglican Clergyman Peter Peckard, proposed an essay competition, asking contributors the question, is it lawful to make slaves of others against their will? (Hochschild 2005, p. 87).

The winner of the essay competition was a young divinity student named Thomas Clarkson who turned out to be one of the most influential members for the antislavery movement. Clarkson was one of the twelve men who met in the printing shop and, according to the minutes of the committee, the meeting was 'held for the purpose of taking the slave trade into consideration and it was resolved that the said trade was both impolitic and unjust' (minutes 22 May 1787, quoted in Hochschild 2005). These men did not envision a perfect society; rather, their campaign was a pragmatic adventure as they were fully aware that it would take great effort to convince people, who perceived slavery as completely normal, that it was wholly unjust. Realizing that much of the profits from the slave trade expanded the economy and was a source of employment for many people, they had to proceed strategically. For instance, one of the first issues pressed upon them was whether they were going to focus on abolishing the slave trade or encourage the emancipation of all slaves. While there were disagreements between the committee members, they agreed to take up the latter and concentrate on fighting one battle at a time.

The *modus operandi* of the committee members was to spread information to the people and to expose the brutality of the slave trade. Clarkson, like a modern-day detective, travelled by horseback to the various slave ports to conduct interviews and gather evidence. Clarkson produced an abstract of the evidence of the slave trade, and according to Hochschild, it reads like a report from a modern-day human rights organization. Committee members circulated pamphlets and copies of the minutes from their meetings. They were good fund raisers as well, attracting influential people who had money and connections. For instance, Josiah Wedgwood had a designer fashion a seal for stamping the wax used to close envelopes. It displayed a kneeling African man in chains with his hands raised and the caption, 'Am I not a man and a brother?' The image was so successful that the antislavery advocate Benjamin Franklin said that it was 'equal to that of the best written pamphlet' (Hochschild 2005, p. 129). Finally, the members of the antislavery committee campaigned for a boycott of sugar. They reasoned that if they informed people of how their sugar was made, then perhaps this would entice them to avoid using it.

The committee members realized that they needed more than a growing following but desperately required a voice in parliament as well. Initially, William Wilberforce, a Member of Parliament, was more concerned with the suppression of sin rather than advocating against slavery, but he turned out to be an essential ally for the committee and became their political authority. Wilberforce introduced the first abolition bill in parliament in 1789, but it was postponed. Persistence paid off, as in 1807 a bill abolishing the slave trade in Britain passed both Houses. Just as has been the case with most progress in the pursuit of 'thick cooperation', this was not a situation where the government provided strong leadership but one 'where the sense of the nation has pressed abolition upon our rulers' (Walvin 1968). Moreover, this struggle to abolish slavery was not accomplished from seeing a divine origin in certain nonnegotiable 'sacred values' but was the result of a bottom-up process *challenging* certain 'sacred values'. In fact, the abolition of slavery was accomplished in spite of the 'sacred values' of the time.

Women's rights

In Chapter 4, it was argued that the struggle in France for human rights by such courageous individuals as Sophie de Grouchy, Marquis de Condorcet and Mary Wollstonecraft paved the way for a less-unjust society and the realization of 'thick cooperation'. Sophie de Grouchy and Condorcet advocated for women's education as they saw this as instrumental in giving women a voice in their private lives, as well as in public affairs. Wollstonecraft's rational approach was not only to identify the inequities suffered by women but provided a voice of reason advocating for the rights of all people. However, we don't need to take our minds back to eighteenth-century Europe to find examples of the struggles for the advancement of justice; similar injustices have been identified quite recently.

For instance, in Canada, prior to the Charter of Rights and Freedoms, husbands could not be charged criminally with raping their wives. The belief that a man's home is his castle and within that home he could do as he wished was a type of 'sacred value' embraced by most people in society. For the majority of people, this included the man's prerogative to have sex with his wife, regardless of whether she consented or not. The argument here is that if a particular society does not pursue equality or allow certain 'cherished values' to be challenged and allow all members of society the freedom to reason against nonnegotiable values, then 'thick cooperation' is nothing but an illusion. Unfortunately, there are still countries in the world that have antiquated laws justified by their 'sacred values', and it is incomprehensible as to how a transcendental approach of values or justice as offered by the BGP could make those societies a little less unjust.

The road to 'Denmark'

Ironically, the road to 'Denmark' has been built from not clearly knowing what 'Denmark' is or looks like. What has been important, and certainly will be important in the future, is to stay clear from envisioning a just society with nonnegotiable values as the blueprint for greater cooperation. Rather, we must be brave enough to challenge our 'sacred values' to make the world a little less unjust for others. This seems to be the only way to achieve 'thick cooperation'.

References

Acemoglu, Daron and James Robinson. (2012). *Why nations fail.* New York: Crown Publishers.

Ahmed, Ali and Mats Hammarstedt. (2011). 'The effect of subtle religious representations on cooperation.' *International Journal of Social Economics* 38: 900–910.

Ahmed, Ali and Osvaldo Salas. (2008). 'In the back of your mind: Subliminal influences of religious concepts on prosocial behaviour.' *Gothenburg School of Business, Economics and Law (Working Papers).* Gothenburg, Sweden: University of Gothenburg.

Ahmed, Ali and Osvaldo Salas. (2011). 'Implicit influences of Christian religious representations on dictator and prisoner's dilemma game decisions.' *Journal of Socio-Economics* 40: 242–246.

Aiello, Leslie and Peter Wheeler. (1995). 'The expensive-tissue hypothesis: The brain and the digestive system in human and primate evolution.' *Current Anthropology* 36(2): 199–221.

Alexander, Richard. (1987). *The biology of moral systems.* New York: Aldine De Gruyter.

Allman, John. (2000). *Evolving brains.* New York: W.H. Freeman and Company.

Almenberg, Johan and Anna Dreber. (2013). 'Economics and evolution: Complementary perspectives on cooperation.' In Martin Nowak and Sarah Coakley (Eds.), *Evolution, games and God: The principle of cooperation* (132–149). Cambridge: Harvard University Press.

Anderson, Benedict. (1983). *Imagined communities: Reflections on the origin and spread of nationalism.* London: Verso.

Asch, Solomon. (1956). 'Studies of independence and conformity: A minority of one against a unanimous majority.' *Psychological Monographs* 70(9): 1–70.

Atran, Scott. (2002). *In gods we trust: The evolutionary landscape of religion.* Oxford: Oxford University Press.

Atran, Scott. (1990). *Cognitive foundations of natural history. Towards an anthropology of science.* Cambridge: Cambridge University Press.

Axelrod, Robert. (1984). *The evolution of cooperation.* New York: Perseus Books Group.

Bargh, John and Tanya Chartrand. (1999). 'The chameleon effect: The perception behaviour link and social interaction.' *Journal of Personality and Social Psychology* 76: 893–910.

Barrett, Justin. (2004). *Why would anyone believe in God?* New York: Altamira Press.

Bateson, Melissa, Daniel Nettle and Gilbert Roberts. (2006). 'Cues of being watched enhance cooperation in a real-world setting.' *Biology Letters* 2(3): 412–414.

Batson, Daniel and Janine Flory. (1990). 'Goal-relevant cognitions associated with helping by individuals high on intrinsic, and religion.' *Journal for the Scientific Study of Religion* 29(3): 346–360.

Batson, Daniel and Tecia Moran. (1999). 'Empathy-induced altruism in a prisoner's dilemma'. *European Journal of Social Psychology* 29(7): 909–924.

Bhalla, Mukul and Dennis Proffitt. (1999). 'Visual-motor recalibration in geographical slant perception.' *Journal of Experimental Psychology: Human Perception and Performance* 25: 1076–1096.

Baumeister, Roy, Isabelle Bauer and Stuart Lloyd. (2010). 'Choice, free will, and religion.' *Psychology of Religion and Spirituality* 2(2): 67–82.

Belisle, Patrick and Bernard Chapais. (2001). 'Tolerated co-feeding in relation to degree of kinship in Japanese macaques.' *Behaviour* 138: 487–509.

Bellah, Robert. (2011). *Religion in human evolution*. Cambridge: Belknap Press.

Bentham, Jeremy. (1843). 'Anarchical Fallacies: Being an examination of the Declaration of Rights issued during the French Revolution.' In John Bowring (ed.) The works of Jeremy Bentham, volume 2. Edinburgh: William Tait. Accessed from http://oll.libertyfund.org/title/1921/114226.

Bering, Jesse. (2007). 'Science will never silence God.' In John Brockton (Ed.), *What is your dangerous idea? Today's leading thinkers on the unthinkable* (167–168). New York: Pocket Books.

Bering, Jesse and Dominic Johnson. (2006). 'Hand of god, mind of man: Punishment and cognition in the evolution of cooperation.' *Human Nature* 4: 219–233.

Bering, Jesse and Dominic Johnson. (2005). 'O Lord you perceive my thoughts from afar: Recursiveness and the evolution of supernatural agency.' *Journal of Cognition and Culture* 5: 118–142.

Berman, Carol. (2016). 'Primate kinship: Contributions from Cayo Santiago.' *American Journal of Primatology* 78: 67–77.

Berna, Francesco, Paul Goldberg, Lora Howrwitz, James Brink, Sharon Holt, Marion Bamford and Michael Chazan. (2012). 'Microstratigraphic evidence of in situ fire in the acheulean strata of Wonderwerk cave, Northern Cape Province, South Africa.' *Proceedings of the National Academy of Sciences* 109(20): 1215–1220.

Bevc, Irene and Irwin Silverman. (2000). 'Early separation and sibling incest: A test of the revised Westermarck theory.' *Evolution and Human Behaviour* 21: 151–161.

Bickerton, Derek. (2014). *More than nature needs*. London: Harvard University Press.

Bickerton, Derek. (2009). *Adam's tongue: How humans made language and how language made humans*. New York: Farrar, Straus and Giroux.

Boehm, Christopher. (2012). *Moral origins*. New York: Basic Books.

Boehm, Christopher. (1997). 'Impact of the human egalitarian syndrome on Darwinian selection mechanics.' *American Naturalist* 150: 100–121.

Bohnet, Daniel, Kathryn Oleson, Joy Weeks, Sean Healy, Penny Reeves and Patrick Jennings. (1989). 'Religious prosocial motivation: Is it altruistic or egoistic?' *Journal of Personality and Social Psychology* 57(5): 873–884.

Bohnet, Iris and Bruno Frey. (1999). 'The sounds of science in prisoner's dilemma and dictator's games.' *Journal of Economic Behaviour and Organization* 38(1): 43–57

Bohnet, Iris, Bruno Frey and Steffen Huck. (2001). 'More order with less law: On contract enforcement, trust and crowding.' *American Political Science Review* 95: 131–144.

Booth, Cathy. (2001). 'The bad samaritan.' Los Angeles: *The L.A. Times*. http://content.time.com/time/magazine/article/0,9171,139892,00.html.

Boyd, Robert and Peter Richardson. (1992). 'Punishment allows the evolution of cooperation (or anything else) in sizeable groups.' *Ethology and Sociobiology* 13: 171–195.

Boyer, Pascal. (2001). *Religion explained*. New York: Basic Books.

Boyer, Pascal. (1998). 'Cognitive tracks of cultural inheritance: How evolved intuitive ontology governs cultural transmission.' *American Anthropologist* 100: 876–889.

Boyer, Pascal. (1994). 'Cognitive constraints on cultural representations: Natural ontologies and religious ideas'. In Lawrence Hirschfield and Susan Gelman (Eds.), *Mapping the mind: Domain-specificity in culture and cognition* (391–411). New York: Cambridge University Press.

Bowles, Samuel and Herbert Gintis. (2011). *A cooperative species*. Princeton: Princeton University Press.

Brauer, Julian, Josep Call and Michael Tomasello. (2006). 'Are apes really inequity averse?' *Proceedings of the Royal Society B, Biological Sciences* 273(1605): 3123–3128.

Brauer, Juliane and Daniel Hanus. (2012). 'Fairness in non-human primates?' *Social Justice Research* 25: 256–276.

Brosnan, Sarah and Frans de Waal. (2003). 'Monkeys reject unequal pay.' *Nature* 425: 297–299.

Brosnan, Sarah and Michael Beran. (2009). 'Trading behaviour between conspecifics in chimpanzees, Pan Troglodytes.' *Journal of Comparative Psychology* 123(2): 181–194.

Brosnan, Sarah and Frans de Waal. (2014). 'Evolution of responses to (un)fairness.' *Science* 346(6207): 1251776-1–1251776-7.

Brownell, Celia, Margartita Svetlova and Sara Nichols. (2009). 'To share or not to share: When do toddlers respond to another's needs?' *Infancy* 14(1): 117–130.

Buckley, Cara. (2007, Jan 7). 'Why our hero leapt onto the tracks and we might not'. *New York Times*.

Brownell, Celia. (2011). 'Early developments in joint action.' *Review of Philosophy and Psychology* 2: 193–211.

Burkart, Judith, Charles Jefferson, Ernst Fehr and Carol Van Schaik. (2007). 'Other-regarding preferences in a non-human primate: Common marmosets provision food altruistically.' *Proceedings of the National Academy of Sciences of the United States of America* 104(50): 19762–19766.

Cacioppo, John and William Patrick. (2008). *Loneliness: Human nature and the need for human connection*. New York: W.W. Norton and Company Inc.

Cameron, Judy and David Pierce. (1994). 'Reinforcement, reward, and intrinsic motivation: A meta-analysis.' *Review of Educational Research* 64: 363–423.

Callan, Mitchell, Aaron Kay, James Olsen, Novjyot Bear and Nicole Whitefield. (2010). 'The effects of priming legal concepts on perceived trust and competitiveness, self interested attitudes and competitive behaviour.' *Journal of Experimental Social Psychology* 46(2): 325–335.

Carneiro, Robert. (1970). 'A theory of the origin of the state.' *Science* 169: 733–738.

Cesario, Joseph. (2014). 'Priming, replication and the hardest science.' *Perspectives on Psychological Science* 9: 40–48.

Chabris, Christopher and Daniel Simons. (2009). *The invisible gorilla: How our intuitions deceive us*. New York: Random House Inc.

Chapais, Bernard. (2011). 'The deep social structure of humankind.' *Science* 331: 1276–1277.

Chapais, Bernard. (2008). *Primeval kinship: How pair-bonding gave birth to human society*. Cambridge: Harvard University Press.

Chaudhuri, Ananish. (2011). 'Sustaining cooperation in laboratory public goods experiments: A selective survey of the literature.' *Economic Science* 14: 47–83.

Cheney, Dorothy and Robert Seyfarth. (1989). 'Redirected aggression and reconciliation among vervet monkeys, *Cercopithecus aethiops*.' *Behaviour* 110: 258–275.

Cheney, Dorothy and Robert Seyfarth. (1986). 'The recognition of social alliances by vervet monkeys.' *Animal Behaviour* 34: 1722–1731.

Chernyak, Nadia and Tamar Kushnir. (2014). 'The self as a moral agent: Preschoolers behave morally but believe in the freedom to do otherwise.' *Journal of Cognition and Development* 15: 1343–1355.

Churchland, Patricia. (2013). *Touching a nerve*. New York: W.W. Norton and Company.

Churchland, Patricia. (2011). *Braintrust*. Princeton: Princeton University Press.

Cialdini, Robert and Noah Goldstein. (2004). 'Social influence: Compliance and conformity.' *Annual Review of Psychology* 55(1): 591–621.

Claidiere, Nicholas, Thomas Phillips and Dan Sperber. (2014). 'How darwinian is cultural evolution?' *Biological Science* 369 (1642): 1–8.

Clutton-Brock, Tim. (2009). 'Cooperation between non-kin in animal societies.' *Nature* 462: 51–57.

Coelho, Philip and James McClure. (2016). 'The evolution of human cooperation.' *Journal of Bioecon* 18: 65–78.

Cosmides, Leda and John Tooby. (1992). 'Cognitive adaptations for social exchange.' In Jerome Barkow, Leda Cosmides and John Tooby (Eds.), *The adaptive mind* (163–228). New York: Oxford University Press.

Crisp, Roger. (2014). *Translation of Nicomachean ethics by Aristotle*. Cambridge: Cambridge University Press.

Csikszentmihalyi, Mihaly. (1990). *Flow*. New York: Harper and Row.

Curley, James and Eric Keverne. (2005). 'Genes, brains and mammalian social bonds.' *Trends in Ecology and Evolution* 20(10): 561–567.

Currier, Richard. (2017). *Unbound: How eight technologies made us human, transformed society, and brought our world to the brink*. New York: Arcade Publishing.

Curry, Oliver and Robin Dunbar. (2013a). 'Do birds of a feather flock together? The relationship between similarity and altruism in social networks.' *Human Nature* 24: 336–347.

Curry, Oliver, Scott Roberts and Robin Dunbar. (2013b). 'Altruism in social networks: Evidence for a "kinship premium".' *British Journal of Psychology* 104: 283–295.

Cutler, Brian and Steven Penrod. (1995). 'Choosing, confidence and accuracy: A meta-analysis of the confidence-accuracy relation in eyewitness identification studies.' *Psychological Bulletin* 118: 315–327.

D'Costa, Gavin. (2005). *Theology in the public square*. Oxford: Blackwell Publishing.

Damasio, Antonio. (2010). *Self comes to mind*. New York: Vintage Books.

Darwin, Charles. (2004/1879). *The descent of man*. London: Penguin Books.

Darwin, Charles. (1999/1859). *The origin of species*. New York: Bantam Books.

Datta, Ansua. (2017). 'California's three strikes law revisited: Assessing the long term effects of the law.' *Atlantic Economic Journal* 45(2): 225–249.

Dawkins, Richard. (1989). *The selfish gene*. London: Oxford University Press.

Dawkins, Richard. (1982). *The extended phenotype*. London: Oxford University Press.

Dawkins, Richard. (2006). *The God delusion*. New York: Houghton Mifflin Company.

Deacon, Terrence. (1997). *The symbolic species*. New York: W.W. Norton Company.

Deci, Edward, Richard Koestner and Richard Ryan. (2001). 'Extrinsic rewards and intrinsic motivation in education: Reconsidered once again.' *Review of Educational Research* 71: 1–27.

Deci, Edward. (1971). 'Effects of externally mediated rewards on intrinsic motivation.' *Journal of Personality and Social Psychology* 18: 105–115.

Dehaene, Stanislas, Lionel Naccache, Jean-Pierre Change, Jerome Sackur and Clair Sergent. (2006). 'Conscious, preconscious and subliminal processing: A testable taxonomy.' *Trends in Cognitive Sciences* 10: 204–211.

de Waal, Frans. (2013). *The bonobo and the atheist*. New York: W.W. Norton and Company.

de Waal, Frans. (1989). 'Food sharing and reciprocal obligation in chimpanzees.' *Journal of Human Evolution* 18: 433–459.

Dennett, Daniel. (2017). *From bacteria to Bach and back: The evolution of minds*. New York: W.W. Norton and Company Inc.

Dennett, Daniel (2013). *Intuition Pumps and other tools for thinking*. New York: W.W. Norton & Company.

Dennett, Daniel. (2006). *Breaking the spell*. New York: Viking Books.

Dennett, Daniel. (1995). *Darwin's dangerous idea*. New York: Touchstone Books.

Diamond, Jared. (1997). *Guns, germs and steel*. London: Vintage Books.

Dijksterhuis, Ap, Russel Spears, Tom Postmes, Diederik Stapel, Willem Koomen, Ad Van Knippenberg and Daan Scheepers. (1998). 'Seeing one thing and doing another: Contrast effects in automatic behaviour.' *Journal of Personality and Social Psychology* 75: 862–871.

Doyen, Stephanie, Daniel Simons, Axel Cleeremans and Oliver Klein. (2014). 'On the other side of the mirror: Priming in cognitive and social psychology.' *Social Cognition* 32: 12–32.

Doyen, Stephanie, Oliver Klein, Cora-Lisa Pichon and Axel Cleeremans. (2012). 'Behavioural priming: It's all in the mind, but who's mind?' *Plus One* 7: 1–7.

Drago Francesco, Roberto Gabiati and Pietro Vertova. (2011). 'Prison conditions and recidivism.' *America Law and Economics Review* 13(1): 103–130.

Dunbar, Robin. (2016). *Human evolution*. Oxford: Oxford University Press.

Dunbar, Robin. (1996). *Grooming, gossip and the evolution of language*. Cambridge: Harvard University Press.

Dunn, John. (2005). *Democracy: A history*. Toronto: Penguin Books.

Dunfield, Kristen and Valerie Kuhlmeier. (2010). 'Intention-mediated selective helping in infancy.' *Psychological Science* 21(4): 523–527.

Durkheim, Emile. (2001/1912). *The elementary forms of religious life*. Oxford: Oxford World Classics.

Edelman, Gerald. (2004). *Wider than the sky: The phenomenal gift of consciousness*. New Haven: Yale University Press.

Edgell, Penny, Joseph Gerteis and Douglas Hartmann. (2006). 'Atheists as other: Moral boundaries and cultural membership in American society.' *American Sociological Review* 71(2): 211–234.

Eliade, Mircea. (1964). *Archaic techniques of ecstasy*. New York: Pantheon Books.

Erdal, David and Andrew Whiten. (1994). 'On human egalitarianism: An evolutionary product of Machiavellian status escalation.' *Current Anthropology* 35: 175–184.

Evans, Jonathan. (2010). *In two minds: Dual processes and beyond*. Oxford: Oxford University Press.

Evans, Jonathan. (2008). 'Dual processing accounts of reasoning judgment and social cognition.' *Annual Review of Psychology* 59: 255–278.

Falk, Dean. (2004). 'Prelinguistic evolution in early hominids: Whence mothers?' *Behavioural and Brain Sciences* 27(4): 491–503.

Fehr, Ernst and Urs Fischbach. (2003). 'The nature of human altruism.' *Nature* 425(6960): 785–791.

Fehr, Ernst and Simon Gachter. (2002). 'Altruistic punishment in humans.' *Nature* 415(6868): 137–140.

Finley, M.I. (1959). 'Was Greek civilization based on slave labour?' *History Zeitschrift fur alte Geschichte* 8: 145–164.

Fischer, Klaus, Susann Janowitz, Stefan Dotter, Stephanie Heuskin and Isabell Karl. (2015). 'Kin recognition and inbreeding avoidance in a butterfly.' *Ethology* 121(10): 977–984.

Frank, Robert. (1988). *Passions within reason*. New York: W.W. Norton and Company.

Fukuyama, Francis. (2011). *The origin of political order*. New York: Farrar, Straus and Giroux.

Galen, Luke. (2012). 'Does religious belief promote pro-sociality?' *Psychological Bulletin* 138(5): 876–906.

Galinsky, Adam, Gillian Ku and Cynthia Wong. (2008). 'Perspective-takers behave more stereotypically.' *Journal of Personality and Social Psychology* 95(2): 409–419.

Gamble Clive, Robin Dunbar and John Gowlett. (2014). *Thinking big: How the evolution of social life shaped the human mind*. London: Thames and Hudson.

Gat, Azar. (2013). *Nations*. Cambridge: Cambridge University Press.

Geertz, Clifford. (1973). 'Thick description: Toward an interpretive theory of culture'. In *The interpretation of cultures: Selected essays* (3–30). New York: Basic Books.

Gervais, Will and Ara Norenzayan. (2012). 'Like a camera in the sky? Thinking about God increases public self-awareness and socially desirable responding.' *Journal of Experimental Social Psychology* 48: 298–302.

Gintis, Howard, Samuel Bowles, Robert Boyd and Ernst Fehr. (2003). 'Explaining altruistic behaviour in humans.' *Evolution and Human Behaviour* 24(3): 153–172.

Godson, James and Andrew Bass. (2001). 'Social behaviour functions and related anatomical characteristics of vasotocin/vasopressin systems in vertebrates.' *Brain Research Reviews* 35(3): 246–265.

Gomes, Christina and Michael McCullough. (2015). 'The effects of implicit primes on dictator game allocations: A preregistered replication experiment.' *Journal of Experimental Psychology* 144(6): 94–104.

Goodall, Jane. (2006). *The Chimpanzees of Gombe: Patterns of Behaviour*. Cambridge: Harvard University Press.

Gopnik, Alison, Andrew Meltzoff and Patricia Kuhl. (1999). *The scientist in the crib*. New York: Harper Publishers.

Gough, Kathleen. (1971). 'The origin of the family.' *Journal of Marriage and the Family* 33: 760–771.

Gray, James. (2001). *Why our drug law have failed and what we can do about it*. Philadelphia: Temple University Press.

Greene, Joshua. (2013). *Moral tribes*. New York: The Penguin Press.

Greenfield, Patricia. (1991). 'Language, tools and brain: The ontogeny and phylogeny of hierarchically organized sequential behaviour.' *Behaviour and Brain Science* 14(4): 531–595.

Griffin, Ashleigh and Stuart West. (2003). 'Kin discrimination and the benefit of helping in cooperatively breeding vertebrates.' *Science* 302(5645): 634–636.

Haidt, Jonathon. (2012). *The righteous mind*. New York: Vintage Books.

Haidt, Jonathon. (2001). 'The emotional dog and its rational tail.' *Psychological Review* 108: 814–834.

Hamann, Katharina, Felix Warneken, Julia Greenberg and Michael Tomasello. (2011). 'Collaboration encourages equal sharing in children but not in chimpanzees.' *Nature* 476(7360): 328–331.

Hamilton, William. (1964). 'The genetical evolution of social behaviour.' *Journal of Theoretical Biology* 7(1): 1–52.

Harcourt, Bernard. (2001). *The illusion of order: The false promises of the broken windows policing*. Cambridge: Harvard University Press.

Hardecker, Susanne, Marco Schmidt and Michael Tomasello. (2017). 'Children's developing understanding of the conventionality of rules.' *Journal of Cognition and Development* 18(2): 163–188.

Hare, Brian, Alicia Melis and Michael Tomasello. (2006). 'Engineering cooperation in chimpanzees: Tolerance constraints on cooperation.' *Animal Behaviour* 72(2): 275–286.

Harris, Sam. (2014). *Waking up: A guide to spirituality without religion*. New York: Simon and Schuster.

Haun, Daniel, Michael Tomasello and Yvonne Rekers. (2014). 'Children conform to the behaviour of peers; other great apes stick with what they know.' *Psychological Science* 25(12): 2160–2167.

Haun, Daniel and Michael Tomasello. (2011). 'Conformity to peer pressure in preschool child.' *Child Development* 82: 1759–1767.

Hayden, Brian. (1995). 'Pathways to power: Principles for creating socioeconomic inequalities.' In Douglas Price and Gary Feinman (Eds.), *Foundations of social inequality* (15–86). New York: Plenum Press.

Hays-Gilpin, Kelley. (2004). *Ambiguous images: Gender and rock art*. Walnut Creek: Atamira Press.

Helwig, Charles and Elliot Turiel. (2017). 'The psychology of children's rights.' In Martin Ruck and Michele Peterson-Badali (Eds.), *Handbook of children's rights: Global and multidisciplinary perspectives* (In press). New York: Rutledge.

Henrich, Joseph and Francisco Gil-White. (2001). 'The evolution of prestige freely conferred deference as a mechanism for enhancing the benefits of cultural transmission.' *Evolution and Human Behaviour* 22: 165–196.

Henrich, Joseph. (2016). *The secret of our success*. Princeton: Princeton University Press.

Hepach, Robert, Katharina Haberl, Stephanie Lambert and Michael Tomasello. (2017). 'Toddlers help anonymously.' *Infancy* 22(1): 130–145.

Hepach, Robert, Amrisha Vaish and Michael Tomasello. (2013). 'Young children sympathize less in response to unjustified emotional distress.' *Developmental Psychology* 49: 1132–1138.

Hepach, Robert, Amrisha Vaish and Michael Tomasello. (2012). 'Young children are intrinsically motivated to see others helped.' *Psychological Science* 23(9): 967–972.

Herculano-Houzel, Suzana. (2016). *The human advantage: How our brains became remarkable*. Cambridge: MIT Press.

Hewlett, Barry. (1991). *Intimate fathers: The nature and context of Aka Pygmy paternal infant care*. Ann Arbour: University of Michigan Press.

Hill, Kim. (2002). 'Altruistic cooperation during foraging by the Ache, and the evolved human predisposition to cooperate.' *Human Nature* 13(l): 105–128.

Hitchens, Christopher. (2007). *God is not great*. Toronto: McClelland and Stewart.

Hobbes, Thomas. (1996/1651). *Leviathan*. New York: Oxford University Press.

Hochschild, Adam. (2005). *Bury the chains*. New York: Houghton Mifflin Company.

Hodges, Andrew. (1983). *Alan Turing: The enigma*. New York: Simon and Schuster.

Hogarth, Robin. (2014). 'Automatic processes, emotions and the causal field.' *Behavioural and Brain Sciences* 37: 1–61.

Holtgraves, Thomas. (1989). 'Conversational memory: The effects of speaker status on memory for the assertiveness of conversational remarks.' *Journal of Personality and Social Psychology* 56: 149–160.

Hood, Bruce. (2009). *The self illusion: How the social brain creates identity*. Toronto: HarperCollins Publishers Ltd.

Hrdy, Sarah Blaffer. (2009). *Mothers and others: The evolutionary origins of mutual understanding*. Cambridge: Belknap Press of Harvard University Press.

Hrdy, Sarah Blaffer. (2005). 'Comes the child before man: How cooperative breeding and prolonged post weaning dependence shaped human potential.' In Barry Hewlett and Michael Lamb (Eds.), *Hunter-gatherer childhoods* (65–91). New Jersey: Transaction Publishers.

Humphrey, Nicholas. (1995). *Social searching: Human nature and supernatural belief*. London: Chatto and Windus.

Hunt, Lynn. (2007). *Inventing human rights*. New York: W.W. Norton and Company Inc.

Hunt, Lynn. (1996). *The French revolution and human rights*. Boston: Bedford Books.

Insel, Thomas. (2010). 'The challenge of translation in social neuroscience: A review of oxytocin, vasopressin and affiliative behaviour.' *Neuron* 65: 768–779.

Ivey, Paula. (2000). 'Cooperative reproduction in Ituri forest hunter-gatherers: Who cares for Efe infants?' *Current Anthropology* 41(5): 856–866.

James, William. (1981/1890). *The principles of psychology*. Cambridge: Harvard University Press.

James, William. (1997/1902). *The varieties of religious experience*. New York: Touchstone Books.

Jiang, J.J., Gary Klein and Richard Vedder. (2000). 'Persuasive expert systems: The influence of confidence and discrepancy.' *Computers in Human Behaviour* 16(2): 99–109.

Johanson, Donald and Blake Edgar. (1996). *From Lucy to language*. New York: Simon and Schuster.

Johnson, Allen and Timothy Earl. (2000). *The evolution of human societies*. Stanford: Stanford University Press.

Johnson, Dominic. (2016). *God is watching you: How the fear of God makes us human*. Oxford: Oxford University Press.

Johnson, Dominic and Oliver Kruger. (2004). 'The good of wrath: Supernatural punishment and the evolution of cooperation.' *Political Theology* 5(2): 159–176.

Johnson, Kathryn, Adam Cohen and Morris Okun. (2015). 'God is watching you ....but also watching over you: The influence of benevolent God representation on secular volunteerism among Christians.' *Psychology of Religion and Spirituality* 1–12.

Johnson, Kathryn, Morris Okun and Adam Cohen. (2014). 'The mind of the Lord: The development of a scale to assess benevolent and authoritarian God-concepts.' *Psychology of Religion and Spirituality* 7(3): 227–238. Advance online publication.

Johnson, Megan, Wade Rowatt and Jordan LaBouff. (2010). 'Priming Christian religious concepts increases racial prejudice.' *Social Psychological & Personality Science* 1: 119–126.

Kahneman, Daniel. (2011). *Thinking fast and slow*. New York: Doubleday.

Kanwisher, Nancy, Josh McDermont and Marvin Chun. (1997). 'The fusiform face area: A module in human extrastriate cortex specialized for face perception.' *The Journal of Neuroscience* 17: 4302–4311.

Kaplan, Hillard, Magdalena Hurtado, Jane Lancaster and Kim Hill. (2000). 'A theory of human life history evolution: Diet, intelligence and longevity.' *Evolutionary Anthropology* 9: 156–185.

Kay, Aaron, John Bargh, Christian Wheeler and Lee Ross. (2004). 'The influence of mundane physical objects on situations construal and competition behavioural choice.' *Organizational Behaviour and Human Decision Processes* 95: 83–96.

Keil, Frank. (1994). 'The birth and nurturance of concepts by domains: The origins of concepts of living things.' In Lawrence Hirschfeld and Susan Gelman (Eds.), *Mapping the mind: Domain specificity in cognition* (234–254). New York: Cambridge University Press.

Kelly, John and Ansua Datta. (2009). 'Does three strikes really deter? A statistical analysis of its impact on crime rates in California.' *College Teaching Methods and Style Journal* 5(1): 29–36.

Kelly, Raymond. (2005). 'The evolution of lethal intergroup violence.' *Proceedings of the National Academy of Sciences*. 102: 15294–98.

Keltner, Dacher and Brenda Buswell. (1996). 'Evidence for the distinctiveness of embarrassment, shame and guilt: A study of recalled antecedents and facial expressions of emotion.' *Cognition and Emotion* 10(2): 155–171.

Kiehl, Kent. (2014). *The psychopath whisperer*. New York: Crown Publishers.

Kihlstrom, John. (2004). 'Is there a people are stupid school in social psychology?' *Behavioural and Brain Sciences* 27(3): 348.

Killen, Melanie and Adam Rutland. (2013). *Children and social exclusion*. Oxford: Blackwell Publishing Ltd.

Konner, Melvin. (2010). *The evolution of childhood*. Cambridge: Belknap Press of Harvard University Press.

Konner, Melvin. (2005). 'Hunter-gatherer infancy and childhood.' In Barry Hewlett and Michael Lamb (Eds.), *Hunter-gatherer childhoods* (123–159). New Jersey: Transaction Publishers.

Kosslyn, Stephen. (2013). 'Social prosthetic systems and human motivation: One reason why cooperation is fundamentally human.' In Martin Nowak and Sarah Coakley (Eds.), *Evolution, Games and God* (153–167). London: Harvard University Press.

Krebs, John and Richard Dawkins. (1984). 'Animal signals: Mind-reading and manipulation.' In John Krebs and Nicholas Davies (Eds.), *Behavioural ecology: An evolutionary approach* (188–205). Oxford: Blackwell Press.

Kroll, Yoraim and Haim Levy. (1992). 'Further tests of the separation theorem and the capital Asset pricing model.' *American Economic Review* 82: 664–670.

Kundt, Radek. (2015). *Contemporary evolutionary theories of culture and the study of religion*. London: Bloomsbury Academic.

Laland, Kevin. (2017). *Darwin's unfinished symphony*. Princeton: Princeton University Press.

Lattimore, Owen. (1962). 'On the wickedness of being nomads.' In *Studies in frontier history: Collected papers 1928–1958* (415–426). London: Oxford University Press.

Laurin, Kristin, Aaron Kay and Grainne Fitzsimons. (2011). 'Divergent effects of activating thoughts of God on self-regulation.' *Journal of Personality and Social Psychology* 102(1): 4–21.

Lawson, E.Thomas and Robert McCauley. (1990). *Rethinking religion: Connecting cognition and culture*. Cambridge: Cambridge University Press.

Lederman, Leon and Dick Teresi. (2006). *The God particle*. New York: Mariner Books.

Levi-Strauss, Claude. (1969). *The elementary structures of kinship*. London: Eyre and Spottiswoode.

Lewis-Williams, David. (2010). *Conceiving God: The cognitive origin and evolution of religion*. London: Thames and Hudson.

Lewis-Williams, David. (2002). *The mind in the cave*. London: Thames and Hudson.

Lieberman, Matthew. (2007). 'Social cognitive neuroscience: A review of core processes.' *Annual Review of Psychology* 58: 259–289.

Liljenquist, Katie, Chen-Bo Zheng and Adam Galinsky. (2010). 'The smell of virtue: Clean scents promote reciprocity and charity.' *Psychological Science* 21(3): 381–383.

Lim, Miranda and Larry Young. (2004). 'Vasopressin-dependent neural circuits underlying pair bond formation in the monogamous prairie vole.' *Neuroscience* 125: 35–45.

MacIntyre, Alasdair. (1981). *After virtue*. Notre Dame: Notre Dame University Press.

Ma-Kellams, Christine and Jim Blascovich. (2013). 'Does 'science' make you moral? The effects of priming science on moral judgments and behavior.' *PLoS One* 8(3): 1–16.

Mann, Michael. (1986). *The sources of social power: A history of power from the beginning to A.D. 1760*. Cambridge: Cambridge University Press.

Martin, Luther and Donald Wiebe. (2014). 'Pro-and assortative-sociality in the formation and maintenance of religious groups.' *Journal for the Cognitive Science of Religion* 2(1): 1–57.

Mate, Gabor. (2008). *In the realm of hungry ghosts*. Toronto: Vintage Canada.

McCall, Cade and Tania Singer. (2012). 'The animal and human neuroendocrinology of social cognition, motivation and behaviour.' *Nature Neuroscience* 15(5): 681–688.

McClenon, James. (2001). *Wondrous healing: Shamans, human evolution and the origin of religion*. Dekalb: Northern Illinois University Press.

McCloskey, Michael, Alfonso Caramazza and Bert Green. (1980). 'Curvilinear motion in the absence of external forces: Naive beliefs about the motion of objects.' *Science* 210: 1139–1141.

McKay, Ryan and Harvey Whitehouse. (2016). 'Religion promotes a love for thy neighbour but how big is the neighbourhood?' *Behavioural and Brain Science* 39: e20.

Melis, Alicia, Patricia Grocke, Josefine Kalbitz and Michael Tomasello. (2016a). 'One for you, one for me, humans unique turn taking skills.' *Psychological Science* 27(7): 987–996.

Melis, Alicia and Felix Warneken. (2016b). 'The psychology of cooperation: Insights from chimpanzees and children.' *Evolutionary Anthropology* 25: 297–305.

Melis, Alicia, Brian Hare and Michael Tomasello. (2008). 'Do chimpanzees reciprocate received favours?' *Animal Behaviour* 76: 951–962.

Miller, Geoffrey. (2001). *The mating mind: How sexual choice shaped the evolution of human nature*. New York: Anchor Books.

Moore, Chris. (2009). 'Fairness in children's resource allocation depends on the recipient.' *Psychological Science* 20(8): 944–948.

Morris, Ian. (2014). *War, what is it good for?* New York: Farrar, Straus and Giroux.

Morris, Ian. (2010). *Why the west rules-for now: The patterns of history and what they reveal about the future*. New York: Farrar, Straus and Giroux.

Murray, Oswyn and Simon Price. (1990). *The Greek city: From Homer to Alexander*. New York: Clarendon Press.

Nagin, Daniel. (2012). 'Crime as pollution.' *American Society of Criminology and Public Policy* 11(2): 275–278.

Nelson, Kevin. (2011). *The spiritual doorway in the brain*. New York: Penguin Group Inc.

Nelson, Leif and Michael Norton. (2005). 'From student to superhero: Situational primes shapes future helping.' *Journal of Experimental Social Psychology* 41(4): 423–430.

Nichols, Shaun and Trisha Folds Bennett. (2003). 'Are children moral objectivists?' *Cognition* 90: b23–b32.

Norenzayan, Ara, Azim Shariff, Aiyana Willard and Teresa Anderson. (2016). 'Religious priming: A meta-analysis with a focus on pro-sociality.' *Personality and Social Psychology Review* 20(1): 27–48.

Norenzayan, Ara, Azim Shariff, Will Gervais, Aiyana Willard, Rita McNamara, Joseph Henrich and Edward Slingerland. (2016a). 'The cultural evolution of prosocial religions.' *Behavioural and Brain Sciences* 39: 1–65.

Norenzayan, Ara. (2013). *Big Gods: How religion transformed cooperation and conflict*. Princeton: Princeton University Press.

North, Douglas, John Wallis and Barry Weingast. (2009). *Violence and social orders: A conceptual framework for interpreting recorded human history*. Cambridge: Cambridge University Press.

Nowak, Martin. (2011). *Super cooperators*. New York: Free Press.

Nowak, Martin and Sarah Cookley (Eds). (2013). *Evolution, games and God*. London: Harvard University Press.

Nozick, Robert. (1974). *Anarchy, state, and utopia*. New Jersey: Basic Books.

Nucci, Larry. (2001). *Education in the moral domain*. Cambridge: Cambridge University Press.

Nucci, Larry and Elsa Weber. (1995). 'Social interactions in the home and the development of young children's conceptions of the personal.' *Child Development* 66: 1438–1452.

Nucci, Larry and Elliot Turiel. (1993). 'God's word, religious rules and their relation to Christian and Jewish childrens' concepts of Morality.' *Child Development* 64: 1475–1491.

Nussbaum, Martha. (2011). *Creating capabilities: The human development approach.* Cambridge: The Belknap Press of Harvard University Press.

Ober, Josiah. (2015). *The rise and fall of classical Greece.* Princeton: Princeton University Press.

Ober, Josiah. (2012). 'Democracy's dignity.' *American Political Science Review* 106(4): 827–842.

Ober, Josiah. (2008). *Democracy and knowledge: Innovation and learning in classical Athens.* Princeton: Princeton University Press.

O'Connell, James, Kristen Hawkes and Nick Blurton Jones. (2002). 'Male strategies and Plio-Pleistocene archeology.' *Journal of Human Evolution* 43: 831–872.

Okun, Morris H. O'Rourke, B. Keller, Kathryn Johnson and C. Enders. (2014). 'Value-expressive volunteer motivation and volunteering by older adults: Relationships with religiosity and spirituality.' *Journal of Gerontology, Series B: Psych Sciences and Social Sciences.*

Packer, Craig. (1977). 'Reciprocal altruism in olive baboons.' *Nature* 265: 441–443.

Pagel, Mark. (2012). *Wired for culture: The natural history of human cooperation.* London: Allen Lang.

Pakosh, Caitlin (Ed.). (2016). *The lawyer's guide to the forensic sciences.* Toronto: Irwin Law Inc.

Panksepp, Jaak and Lucy Biven. (2012). *The archaeology of mind.* New York: W.W. Norton and Co.

Patall, Erika, Harris Cooper and Jorgianne Robinson. (2008). 'The effects of choice on intrinsic motivation and related outcomes: A meta-analysis of research findings.' *Psychological Bulletin* 134: 270–300.

Pentikainen, Juha. (1996). 'Introduction.' In *Shamanism and northern ecology* (1–18). New York: Mouton de Gruyter.

Pfaff, Donald. (2015). *The altruistic brain.* New York: Oxford University Press.

Piaget, Jean. (1997/1932). *The moral judgment of the child.* New York: Free Press.

Pichon Isabelle, Vassillis Saroglou and Giulio Boccato. (2007). 'Nonconscious influences of religion on pro sociality: A priming study.' *European Journal of Psychology* 37: 1032–1045.

Pinker, Steven. (2012). 'The false allure of group selection.' https://www.edge.org/conversation/the-false-allure-of-group-selection.

Pinker, Steven. (2011). *The better angels of our nature.* London: Viking Press.

Pinker, Steven. (1997). *How the mind works.* New York: W.W. Norton and Company.

Popper, Karl. (1968). *Conjectures and refutations.* New York: Harper and Rowe Inc.

Pritchett, Lant and Michael Woolcock. (2004). 'Solutions when the solution is the problem: Arraying the disarray in development.' *World Development* 32(2): 191–212.

Pruden, Wesley. (2007, Aug). 'ah, there's joy in mudville's precincts.' *The Washington Times.*

Pusey, Anne and Craig Packer. (1987). 'Dispersal and philopatry.' In Barbara Smuts, Dorothy Cheney and Robert Seyfarth (Eds.), *Primate Societies* (250–266). Chicago: University of Chicago Press.

Putnam, Robert and David Campbell. (2010). *American grace: How religion divides and unites us*. New York: Simon and Schuster Paperbacks.

R v. Ewanchuk. (1999). 1 SCR 330, 1999 Supreme Court of Canada.

Rakoczy, Hannes, Felix Warneken and Michael Tomasello. (2008). 'The sources of normativety: Young children's awareness of the normative structure of games.' *Developmental Psychology* 52: 1236–1246.

Randolph-Seng, Brandon and Michal Nielsen. (2007). 'Honesty: One effect of primed religious representations.' *International Journal for the Psychology of Religion* 17: 303–315.

Rapp, Diotima, Jan Engelmann, Esther Herrmann and Michael Tomasello. (2017). 'The impact of choice on young children's prosocial motivation.' *Journal of Experimental Child Psychology* 158: 112–121.

Reddy, Vasudevi. (2008). *How infants know minds*. Cambridge: Harvard University Press.

Reed, David, Jessica Light, Julie Allen and Jeremy Kirchman. (2007). 'Pair of lice lost or parasites regained: The evolutionary history of anthropoid primate lice.' *BMC Biology* 5(7): 1–11.

Richerson, Peter and Robert Boyd. (2005). *Not by genes alone: How culture transformed human evolution*. Chicago: Chicago University Press.

Ridley, Matt. (2010). *The Rational Optimist*. New York: HarperCollins Publishers.

Ritter, Ryan and Jesse Preston. (2013). 'Representation of religious words: Insights for religious priming research.' *Journal for the Scientific Study of Religion* 52(3): 494–507.

Roberts, Gilbert. (2005). 'Cooperation through interdependence.' *Animal Behaviour* 70: 901–908.

Robinson, Paul and John Darley. (2004). 'Does criminal law deter? A behavioural science investigation.' *Oxford Journal of Legal Studies* 24(2): 173–205.

Rodseth, Lars and Shannon Novak. (2006). 'The impact of primatology on the study of human society.' In Jerome Barkow (Ed.), *Missing the revolution: Darwinism for social scientists*, 187–220. New York: Oxford University Press.

Roeder, Oliver, Lauren-Brooke Eisen and Julia Bowling. (2015). 'What caused the crime to decline?' *Brennan Centre for Justice at New York University*. https://www.brennancenter.org/sites/default/files/analysis/Crime_rate_report_web.pdf.

Rosmarin, David, Elizabeth Krumrei and Gerhard Andersson. (2009). 'Religion as a predictor of psychological distress in two religious communities.' *Cognitive Behaviour Therapy* 38: 54–64.

Rossano, Matthew. (2010). *Supernatural selection: How religion evolved*. Oxford: Oxford University Press.

Rousseau, Jean-Jacques. (1968/1762). *The social contract*. Translator: Maurice Cranston. New York: Penguin Books.

Russano, Melissa, Christian Meissner, Fadia Narchet and Saul Kassin. (2005). 'Investigating true and false confessions within a novel experimental paradigm.' *Psychological Science* 16(6): 481–486.

Ryan, Richard and Edward Deci. (1985). *Intrinsic motivation and self-determination in human behaviour*. New York: Plenum Books.

Sahlins, Marshall. (1972). *Stone age economics*. New York: Aldine Publishing.
Sale, Kirkpatrick. (2006). *The evolution of human domination*. London: Duke University Press.
Schumpeter, Joseph. (1942). *Capitalism, socialism and democracy*. New York: Harper Inc.
Scott, James. (2017). *Against the grain: A deep history of the earliest states*. New Haven: Yale University Press.
Seabright, Paul. (2004). *A company of strangers: A natural history of economic life*. Princeton: Princeton University Press.
Searle, John. (2010). *Making the social world. The structure of human civilization*. Oxford: Oxford University Press.
Sen, Amartya. (2009). *The idea of justice*. Cambridge: Belknap Press.
Shanks, David and Ben Newell. (2014). 'Unconscious influences on decision making: A critical review.' *Behavioural and Brain Sciences* 37: 1–61.
Shanks, David, Ben Newell, Eun Lee, Divya Balakrisnan, Lisa Ekelund, Zarius Cenac, Frag Kiski and Christopher Moore. (2013). 'Priming intelligent behaviour: An elusive phenomena.' *PLoS One* 8(4): 1–10.
Shariff, Azim and Ara Norenzayan. (2011). 'Mean gods make good people: Different views of God predict cheater behaviour.' *International Journal for the Psychology of Religion* 21: 85–96.
Shariff, Azim and Ara Norenzayan. (2007). 'God is watching you: Priming god concepts increases prosocial behaviour in an anonymous economic game.' *Psychological Science* 18: 803–809.
Shariff, Azim and Mijke Rhemtulla. (2012). 'Divergent effects of beliefs in heaven and hell on national crime rates.' *PLoS One* 7 (6): 1–5.
Shih, Margaret, Todd Pittinsky and Nalini Ambady. (1999). 'Stereotype susceptibility! Identity salience and shifts in quantitative performance.' *Psychological Science* 10(1): 80–83.
Shils, Edward. (1951). 'The study of the primary groups'. In Daniel Lerner and Harold Fisher (Eds.), *The policy sciences: Recent developments in scope and method* (16–42). Stanford: Stanford University Press.
Silberberg, Alan, Lara Crescimbene, Elsa Addessi, James Anderson and Elisabetta Visalberghi. (2009). 'Does inequity aversion depend on a frustration effect? A test with capuchin monkeys (Cebus apella).' *Animal Cognition* 12: 505–509.
Silk, Joan, Sarah Brosnan, Jennifer Vonk, Joseph Henrich, Daniel Povinelli, Amanda Richardson, Susan Lambeth, Jenny Mascaro and Steven Schapiro. (2005). 'Chimpanzees are indifferent to the welfare of unrelated group members.' *Nature* 437: 1357–1359.
Slingerland, Edward. (2014). *Trying not to try*. New York: Crown Publishers.
Slingerland, Edward, Ara Norenzayan and Joseph Henrich. (2013). 'Religious prosociality: A synthesis.' In Peter Richardson and Morten Christiansen (Eds.), *Cultural evolution: Society, technology, language and religion* (356–378). Cambridge: MIT Press.

Smith, Paula, Claire Goggin and Paul Gendreu. (2002). 'The effects of prison sentences and intermediate sanctions on recidivism: General effects and individual differences'. Ottawa: Ontario Solicitor General of Canada.

Smith, Wilfred Cantwell. (1959). 'Comparative religion: Whither and why?'. In Mircea Eliade and Joseph Kitagawa (Eds.), *The history of religions: Essays in methodology* (31–58). Chicago: Chicago University Press.

Spelke, Elizabeth, Gary Katz, Susan Purcell, Sheryl Ehrlich and Karen Breinlinger. (1994). 'Early knowledge of object motion: Continuity and inertia'. *Cognition* 51: 131–176.

Spohn, Cassia and David Holleran. (2002). 'The effect of imprisonment on recidivism rates of felony offenders: A focus on drug offenders'. *Criminology* 40(2): 329–357.

Stanovich, Keith. (2004). *The robot's rebellion*. Chicago: University of Chicago Press.

Stanovich, Keith. (2011). *Rationality and the reflective mind*. Oxford: Oxford University Press.

Stanovich, Keith and Maggie Topiak. (2012). 'Defining features versus incidental correlates of Type 1 and Type 2 processing'. *Mind Society* 11: 3–13.

Stanovich, Keith. (2015). 'Rational and irrational thought: The thinking that IQ tests miss'. *Scientific American: Mind Special Edition* 23(4): 12–17.

Stepher, Joseph. (1983). *Incest: A biosocial view*. New York: Academic Press.

Stiner, Mary, Ran Barkai and Avi Gopher. (2009). 'Cooperative hunting and meat sharing 400–200 kya at Qesem Cave, Israel'. *Proceedings of the National Academy of Sciences* 106: 13207–13212.

Sun, Lixing. (2013). *The fairness instinct: The Robin Hood mentality and our biological nature*. New York: Prometheus Books.

Svetlova, Margarita, Celia Brownell and Sara Nichols. (2010). 'Toddlers' prosocial behaviour: From instrumental to empathic to altruistic helping'. *Child Development* 81(6): 1814–1827.

Szamado, Szabolcs and Eors Szathmary. (2006). 'Selective scenarios for the emergence of natural language'. *Trends in Ecology and Evolution* 21(10): 555–561.

Tashima, Helen and William Marelich. (1989). 'A comparison of the relative effectiveness of alternative sanctions for DUI offenders'. *California Department of Motor Vehicles* 1, National Academies of science, engineering, medicine.Tomasello, Michael. (2016). *A natural history of human morality*. Cambridge: Harvard University Press.

Tomasello, Michael. (2014). *A natural history of human thinking*. Cambridge: Harvard University Press.

Tomasello, Michael. (2009). *Why we cooperate*. Cambridge: MIT Press.

Tooby, John, Leda Cosmides, Andrew Delton and Max Krasnow. (2015). 'Group cooperation without group selection: Modest punishment can recruit much cooperation'. *PLoS One* 10(4): 1–17.

Trivers, Robert. (1971). 'The evolution of reciprocal altruism'. *The Quarterly Review of Biology* 46: 35–57.

Tuomela, Raimo. (2002). *The philosophy of social practices*. Cambridge: Cambridge University Press.

Turchin, Peter. (2016). *Ultra Society*. Connecticut: Beresta Books

Turiel, Elliot. (2017). 'A psychological perspective on moral reasoning, process of decision-making and moral resistance.' *Contemporary Politics* 23: 19–33.

Turiel, Elliot. (2015). 'Moral development.' In Richard Lerner, Willis Overton and Peter Molenaar (Eds.), *Handbook of child psychology and developmental science, volume 1: theory and method*, 7th edition (484–522). Hoboken: John Wiley & Sons.

Turiel, Elliot. (1983). *The development of social knowledge: Morality and convention*. Cambridge: Cambridge University Press.

Uber, Julia, Katharina Hermann and Michael Tomasello. (2016). 'Extrinsic rewards diminish costly sharing in 3 year olds.' *Child Development* 7(4): 1192–1203.

Vaish, Amrisha, Malinda Carpenter and Michael Tomasello. (2010). 'Young children selectively avoid helping people with harmful intentions.' *Child Development* 81: 1661–1669.

Villettaz, Patrice, Martin Killias and Gwladys Gillieron. (2015). 'The effects of custodial vs non-custodial sanctions on re-offending: An updated systematic review of the state of knowledge'. *The Campbell Library* 11(1): 1–92.

Wade, Lizzy. (2015). 'Birth of moralizing gods.' *Science* 349(6251): 918–922

Walvin, James. (1968). *England, slaves and freedom, 1776–1838*. Jackson: University Press of Mississippi.

Warneken, Felix and Michael Tomasello. (2013a). 'Young children proactively remedy unnoticed accidents.' *Cognition* 126(1): 101–108.

Warneken, Felix and Michael Tomasello. (2013b). 'Parental presence and encouragement do not influence helping in young children.' *Infancy* 18: 345–368.

Warneken, Felix and Michael Tomasello. (2009). 'The roots of human altruism.' *British Journal of Psychology* 100: 455–471.

Warneken, Felix and Michael Tomasello. (2008). 'Extrinsic rewards undermine altruistic tendencies in 20 month olds.' *Developmental Psychology* 44(6): 1785–1788.

Warneken, Felix and Michael Tomasello. (2007). 'Helping and cooperation at 14 months of age.' *Infancy* 11(3): 271–294.

Warneken, Felix and Michael Tomasello. (2006). 'Altruistic helping in human infants and young chimpanzees.' *Science* 3(11): 1301–1303.

Wason, Peter and Philip Johnson-Laird. (1972). *Psychology of reason: Structure and content*. Cambridge: Harvard University Press.

West, Stuart, Ashleigh Griffin and Andy Gardner. (2007a). 'Evolutionary explanation for cooperation.' *Current Biology* 17(16): R661–R672.

West, Stuart, Ashleigh Griffin and Andy Gardner. (2007b). 'Social semantics: Altruism, cooperation, mutualism, strong reciprocity and group selection.' *European Society for Evolutionary Biology* 20: 415–432.

West, Stuart, Clarie El Mouden and Andy Gardner. (2011). 'Sixteen common misconceptions about the evolution of cooperation in human.' *Evolution and Human Behaviour* 32(4): 231–262.

Westermarck, Edward. (1891). *The history of human marriage*. London: MacMillan.
Whitehorn, Penelope, Dave Goulson and Matthew Tinsley. (2009). 'Kin recognition and inbreeding reluctance in bumblebees.' *Apidologie* 40: 627–633.
Whiten, Andrew, Frans de Waal and Victoria Horner. (2005). 'Conformity to cultural norms of tool use in chimpanzees.' *Nature* 437(7059): 737–740.
Wiebe, Donald. (1991). *The irony of theology and the nature of religious thought*. Montreal: McGill-Queen's University Press.
Wilkinson, Gerald. (1984). 'Reciprocal food sharing in the vampire bats.' *Nature* 308: 181–184.
Wilkinson, Gerald and Gerald Carter. (2013). 'Food sharing in vampire bats reciprocal help predicts donations more than relatedness or harassment.' *Biological Sciences* 280(1753): 1–6.
Wilson, David Sloan. (2015). *Does altruism exist? Culture, genes and the welfare of others*. New Haven: Yale University Press.
Wilson, David Sloan. (2002). *Darwin's cathedral*. Chicago: University of Chicago Press.
Winkelman, Michael. (2011). 'Shamanism and the evolutionary origins of spirituality and healing.' *Neuroquantology* 9(1): 54–71.
Winkelman, Michael. (1990). 'Shamans and other 'magico-religious healers': A cross cultural study of their origins, natures and social transformations.' *Ethos* 18(3): 308–352.
Wollstonecraft, Mary. (1995/1792). *A vindication of the rights of man and a vindication of the rights of women*. Sylvania Tomaselli (Ed.). Cambridge: Cambridge University Press.
Wolf, Arthur. (1995). *Sexual attraction and childhood association: A Chinese brief for Edward Westermarck*. Stanford: Stanford University Press.
Wolf, Arthur. (1966). 'Childhood association, sexual attraction and the incest taboo: A Chinese case.' *American Anthropology* 68: 883–898.
Wrangham, Richard. (2009). *Catching fire: How cooking made us human*. New York: Basic Books.
Wrangham, Richard. (1980). 'An ecological model of female-bonded primate groups.' *Behaviour* 75: 262–300.
Wynn, Karen. (2009). 'Constraints on natural altruism.' *British Journal of Psychology* 100(3): 481–485.
Yamamoto, Shinya and Magayuki Tanaka. (2009). 'Do chimpanzees (Pan troglodytes) spontaneously take turns in a reciprocal cooperation task?' *Journal of Comparative Psychology* 123(3): 242–249.
Zhang, Xinmin and Stuart Firestein. (2002). 'The olfactory receptor gene superfamily of the mouse.' *Natural Neuroscience* 5: 124–133.
Zhong, Chen-Bo and Katie Liljenquist. (2006). 'Washing away your sins: Threatened morality and physical cleansing.' *Science* 313(5792): 1451–1452.

Index

abolitionists 205
Acemoglu, Daron 152–3
affinal brotherhoods 126
After Virtue (MacIntyre) 157
agency 11, 25, 26, 68, 118, 181
Aiello, Leslie 104
Aka 123
Alexander, Richard 6, 142, 145
algorithmic mind 22, 23, 26, 46–7
Allman, John 104, 108, 121
Altered State of Consciousness (ASC) 31, 32
American Declaration of Independence 156, 158
Amnesty International 158
Anarchy State and Utopia (Nozick) 96
Anaximander 36, 37
Anaximenes 36, 37
Anderson, Benedict 164
Animal Communication Systems (ASC) 111–12
Ann problem 21
Aristotle 93
assortative sociality 9
Athens 151, 153
Autrey, Wesley 1, 2, 3, 5
Axelrod, Robert 140, 141

baboons 137–8
Barrett, Justin 17, 24, 27, 30, 31, 55
Beccaria, Cesare 166
beneficial behaviour 20, 164
Bentham, Jeremy 157–8
Bering, Jesse 17
Bickerton, Derek 95, 106, 110–16, 118
Big Gods: How Religion Transformed Cooperation and Conflict (Norenzayan) 8
Big Man societies 148
biological adaptation 11, 17, 161
Boehm, Christopher 50–1, 143–4, 150
bonding mechanisms 102

Boyer, Pascal 17, 19, 24–6
Brahmin 160
brain expansion 103, 120
Brosnan, Sarah 154
Burke, Edmund 168
burying the chains 205
Bury the Chains (Hochschild) 205
Byrne, Dick 108

caging 146
Calas, Jean 166
    Affair 166
Campbell, David 77
capuchin monkeys 154
Carneiro, Robert 146
Cash, David 1, 2, 5, 88
Caste System 160, 161, 163
cave art 33–4
Cesaro, Joseph 66
Chapais, Bernard 125–8
Charter of Rights and Freedoms 85, 208
cheater detection device 134, 198
Cheney, Dorothy 107
Church of England 206
Churchland, Patricia 92
circumscription 146
Clarkson, Thomas 206–7
Clutten-Brock, Tim 135
Codrington plantation 206
coercion 5, 52, 130, 132, 148, 150, 151
cognitive misers 21
Cognitive Science of Religion (CSR) 10, 11, 19, 24–5, 56
commitment problems 13, 142–3
competence without comprehension 116
computer tournament 140
conforming 222
constant surveillance 89–90
conventional norms 14, 195–7
Convention on the Elimination of All Forms of Discrimination Against Women (CEDAW) 162

Convention on the Rights of the Child (CRC) 162
conversation-starter 159
conversation-stopper 159–60
cooperation
   collaboration
      in animals 142, 173
      collaborative foraging 114, 183
      definition 4
      mutualistic 135–40
      and sharing 182–6
   enhancement 57
   thick 13, 30, 40, 46, 52, 53, 59, 116, 118, 129, 133, 149, 150, 156, 160, 161, 164, 187, 198, 203, 204, 205, 208
      definition 5
   thin 49, 93, 114, 130, 132, 135, 146, 149, 173, 179, 187, 198, 204
      definition 4–5
cooperative breeding 100, 121–3
counter-dominant tendencies 145
creative destruction 153
crime and punishment 82–9
cultural
   development 4–5
   evolution 42–55, 187–8, 201
   group selection 11, 38, 41, 44–55, 57, 94
Curley, James 102–3

Dalits 161
Damasio, Antonio 97
Darwin, Charles 58, 144
   Darwinian Evolution 42
Darwin's Dangerous Idea (Dennett) 55
Dawkins, Richard 8, 39–40, 42, 46, 97
D'Costa, Gavin 160–4
Declaration of the Rights of Man 167
de Condorcet, Marquis 165, 167, 208
de Grouchy, Sophie 167, 208
democracy 5, 151
'Denmark' 199, 200, 208
Dennett, Daniel 25, 43, 55, 95, 116–18
Deoxyribonucleic Acid (DNA) 15, 40, 62
   Junk 40
de Waal, Frans 154–5
dichotomy thesis 36
Dictator Game (DG) 65
Diderot 165
displacement 111–14, 116
Doctors without Borders 158

Dogon women 150
dominant behaviour 145–6, 149
doorstep condition 150–4
Dresher, Melvin 132
dual-processing cognitive system 10, 40
   System 1 (TASS) 10, 19, 20–1
      and God concepts 24–6
      serves the genes 39, 56
   System 2 10, 19, 21–4
      serves the individual 39, 56
Dunbar, Robin 100–1, 106–9, 128
   Dunbar's number 128
Durkheim, Emile 18

Earle, Timothy 146–7
economy 207
   definition 146–7
   political 146
   subsistent 146
Edelman, Benjamin 89
Efe 123–4
Eliade, Mircea 31
Elite Creation 148–9
Elster, Jon 164
emergence of scientific thought 36–8
energy budgets 103–5
Enlene cave 34
Enlightenment 159–60, 165–6
Erdal, David 145
Evans, Jonathan 23
Evolution of Cooperation, The (Axelrod) 140
Evolution of Reciprocal Altruism, The (Trivers) 133
Expensive Tissue Hypothesis 104–5
experience machine, The 96–7, 98
experimental games 62–3, 71
extractive economic institutions 152–3

factoid 8–9
fair institutions 152–4
fairness instinct 82, 154–6, 183
Fairness Instinct, The (Sun) 154
Falk, Dean 110
Flood, Merril 132
foraging orders 148
Fox, Robin 125
Frank, Robert 13, 51, 142–3, 150
free-floating rationales 55, 116–17
free-riders 6, 48, 50, 96, 144

Gabillou cave 34
Gamble, Clive 100
game theory
    experiments 11, 58, 140, 142, 201
    models 72, 132
Gandhi 161
Gat, Azar 157, 158
Geertz, Clifford 4
Generous Tit for Tat 141–2
God(s)
    a belief in 25–6
    Big God Proponents (BGP)
        definition 9–10
        mean or benevolent 77–89
    concepts 56
    and government 200–2
    Shamism and 31–6
    TASS and 24–6
God Is Watching You (Johnson) 60, 86
Gowlett, John 100
greedy reductionism 95–6
Greek efflorescence 153
Greeks 149–53
Greene, Joshua 20
grooming 106–7, 109, 111
group behaviour in I-mode 115
Guns, Germs and Steel (Diamond) 42

Haidt, Jonathon 28, 30, 56
Hamann, Katharina 183–4
Hamilton, William 5, 95
    Hamilton's rule 95, 101, 125, 131
Harijans 161
Harris, Sam 18
helping and intrinsic reward 179–181
helping behaviour 12, 71, 81, 94–5, 124, 131
    in animals 136–40, 172–3
    in children 169–75
        selective helping 175–9
        intrinsic rewards 179–81
Henrich, Joseph 47
Herculano-Houzel, Suzana 120–1
Hindu ethics 161
Hitchens, Christopher 8
Hobbes, Thomas 2–3, 7, 10, 92, 202, 204
Hochschild, Adam 205
Hollopeter, Cameron 1, 2
Hood, Bruce 26–8, 30
Hrdy, Sarah Blaffer 121–2

human rights 14, 48, 76, 151, 156–68, 189, 194, 200–1
    countering 160–4
    history of 164–8
    idea of 156–8
    movement 205–7
    positive and negative rights 158–60
    women's rights 208
Humphrey, Nicholas 7, 75
Hunt, Lynn 164–7
hunting hypothesis 105–7
Hypersensitive Agent Detection Device (HADD) 26, 31, 86
hypothetical reasoning 21–2, 108

imagined communities 164
imagined empathy 164
'I-mode' 115
inalienable rights 156, 165
incest avoidance 126–7
inclusive economic institutions 152–3
in-group 5, 9, 94, 129, 150, 168, 186, 201, 204
intentional stance 25, 116
intuition pump 55

James, William 17, 18, 170–1
Jefferson, Thomas 165
Johnson, Dominic 60, 65, 67, 82–3, 86, 90, 150
joint commitment 113–16, 171

Kane, Pandurang 160
Kanwar, Roop 162, 163
Keverne, Eric 102–3
Kihlstrom, John 29, 30
King George III 157
kin selection 5, 6, 41, 94–5, 101, 124, 130, 131, 134, 135, 136
kinship premium 100–1
Kishwar, Madhu 162
Konner, Melvin 123–4
Krebs, John 141
Kshatriyas 161
!Kung 123

language 109–14, 118–19
    function of 109
Lascaux cave 34
Levy-Bruhl, Claude 36

Lewis-Williams, David 32–4, 54
Lieberman, Matthew 20
limited access orders 148
local groups 125, 129, 147, 148

Machiavellian hypothesis 107–8
Machiavelli, Niccolo 107
MacIntyre, Alasdair 157
Mailer, Norman 8
mandatory minimum sentences 85
Mann, Michael 146
Martin, Luther 9
May, Robert 141
Milesians 24, 36–8
mindware 22
monotheistic religions 17, 45, 55
Montreal police strike 90
moral
    certitude 30
    conscience 13, 50, 144, 170
    non-moral conscience 176
    principles 14, 169–70
    realism 187–9
    sense 144–5
morality 7, 15, 18, 30, 43, 50, 72, 76, 169, 188, 189
Morris, Ian 148, 199
Muyart, Pierre-Francois 166
mythopoeic way of thinking 36

Nash equilibrium 133
natural rights 157, 165
natural selection 95, 97, 116, 134, 141
natural state 149, 151–3
Near Death Experiences (NDE) 33
Nelson, Kevin 33
Neolithic Revolution 147
Nicomachean Ethics (Aristotle) 93
non-reflective beliefs 24
Norenzayan, Ara 8, 13, 59, 66, 68, 69, 70, 73, 74, 75, 78, 79, 200
North, Douglas 5, 130, 148, 149,150, 151
Nowak, Martin 19, 21, 22, 162, 163, 165, 167, 171, 172
Nozick, Robert 96, 97, 98
Nucci, Larry 29, 190, 191
Nussbaum, Martha 161

Ober, Josiah 152, 153
O'Connell, James 106

Old World monkeys 104, 105, 107
Omydiar, Pierre 90
open access orders 148
open society 79, 151
Origin of Species, The (Darwin) 58
Otterbein, Keith 145
OXFAM 158
oxytocin (OXT) 99, 102, 103

Packer, Craig 137
pair bonding 99, 102, 106, 125
Palaeolithic 32, 33, 54
Panksepp, Jaak 98, 155
payoff matrix 133
Peloponnesian War 151
Pericles 151
Perkins, David 22
Pfaff, Donald 99
Piaget, Jean 169
Pinker, Steven 45, 46
Placebo consolidation 53
political economy 146
Popper, Karl 36, 37
preliminary question 166
priming
    critics 63–5
    experiments 11–12, 30, 63
    and justice terms 70–2
    and mean Gods 77–81
    replication 73–7
prisoner's dilemma (PD) 70, 88, 163, 164
Pritchett, Lant 199
pro-sociality
    proactive 173–5, 209
    reactive 173
Public Good Games (PGG) 65
punishment
    avoidance 57
    Beccaria and 166
    crime and 82–9, 90, 92
    and fairness 81–2
    of free riders 6–7
    helping behaviour without 172, 178, 179, 185, 204
    as social selection 144, 145
    supernatural 8, 10, 57, 59–60, 66, 79, 130, 149–50, 153, 156, 167, 169, 188
Putnam, Robert 77
putting-the-baby-down hypothesis 110

Quakers 206

rapid eye movement (REM) 32
    non-REM sleep 33
    REM sleep 51
Rapoport, Anatol 140
Rapp, Diotima 181
Rational Choice Theory (RCT) 48–9
reciprocal 154
    altruism, 22, 60, 164, 165
    exogamy 125–6
reciprocity 73, 134–7
    direct 6, 130, 135–7, 142
    indirect 6, 130, 142
recruitment 113–14, 160
red colobus monkeys 114
Reddy, Vasudevi 170, 203
reflective mind 20, 22, 23, 24, 68
regional groups 108, 129, 147, 148
religion
    as cultural adaptation 38, 44
    definition 18
    as evolutionary by-product 17
    prosocial 75–6
    religious concepts 68, 195
    religious thinking 17, 36, 86
    roots of 32–6
    as successful group 44–55
remembered present 113
reputation 6, 57, 77, 142, 143, 145, 178, 187
Richerson, Peter 6, 41
Ridley, Matt 34, 40, 41, 43, 140
Roberts, Gilbert 124–5
Robinson, James 152
Rossano, Matt 53–4
Rouffignac cave 33
round dance 112
Rousseau, Jean-Jacques 2, 3
Ryle, Gilbert 4

sacred values 15, 28, 201, 202, 203, 207, 208
    non-negotiable 14, 201, 202, 203, 207, 208
sati 162, 163, 164
scavenging 105, 118, 120
    aggressive 105
    passive 105
Schumpeter, Joseph 153
scientific reasoning 22, 23, 38, 108
Scott, James 52

Searle, John 118–20, 156, 157, 158, 159, 204
secular groups (societies) 13, 54, 59
SEEKING 98–9
selfish gene 39
    theory 40, 47
Selfish Gene, The (Dawkins) 40
Sen, Amartya 49, 158, 164, 187, 202
serial associative cognition 22
Seyfarth, Robert 107
Shamanism 31–6
shared intentionality 12, 114, 115, 129, 131, 138, 203, 204
Shared Intentionality Hypothesis 114
shared intentions 170, 173
Shariff, Azim 66, 68, 69, 70, 73, 74, 75, 78, 79
sharing 4, 112, 139, 147, 181–6
    in animals 136–7, 182–6
Sherrice law 1
Shudras 161
Singh, Maal 162
skyhooks and cranes 55–6, 95
slavery 52, 205, 206, 207
slave trade 206, 207
social animals 96
Social Brain Hypothesis 108–9
social contract 3, 96, 165, 202
Social Domain Theory 29, 189
social selection 144
spirituality 18
stakeholder theory 123, 124–5
standard model (of CSR) 17
Stanovich, Keith 20, 22, 23, 39, 40
status functions 119, 157, 158
still-face experiments 171
strategic information 19, 26, 31, 34
Strauss, Levi 125, 126
strong conformity 186, 187
strong evaluations 188, 190
subsistent economy 146
Sun, Lixing 154
supernatural agents 19, 54, 59, 190
    as biological adaptation 11, 17
    as by-product 10, 23, 24, 31, 41, 57, 188
    definition 15
    discourages selfishness 7, 8, 58, 60, 67, 68
    as functional adaptation 11, 53, 94

holders of strategic information 26
prevent crime 88
Supernatural Monitoring Hypothesis (SMH) 15, 60
Supernatural Punishment Hypothesis (SPH) 83, 182
Supernatural Selection: How Religion Evolved (Rossano) 53
supersense 26–8
surplus creation 146–7
survival machines 39, 40
Swift, Jonathan 58
Szamado, Szabolcs 111
Szathmary, Eors 111

Taylor, Charles 188
Thales 36, 37
Theology in the Public Square (D'Costa) 160
theory of mind (ToM) 25, 57
Thucydides 151
Tit for Tat 141, 142
Tomasello, Michael. 113, 114–16, 150, 171, 175, 186, 204
transcendental institutionalism 202–3
trial by ordeals 59
trichromatic colour vision 104
Trivers, Robert 5, 134, 135, 142, 144, 145
Tronick, Ed. 171
Tryambakayajvan 163
Tuomela, Raimo 115
Turchin, Peter 44–8, 50, 51
Turiel, Elliot 30–1, 46, 189
Turing, Alan 46, 47, 51
turn-taking 138

Ultra Society (Turchin) 48

Universal Declaration of Human Rights (UDHR) 157–9
universal rights 159
Untouchables 161
utilitarianism 157

Vaisyas 161
vampire bats 136
Vanita, Ruth 162
Vasotocin 99, 102
Voltaire 166
von Neumann, John 132

waggle dance 112
Wallis, John 5, 130
Warneken, Felix 171, 174, 175
Wason Four Card Selection Task 22–3
weak evaluations 188, 190
Webb, Richard 43
Weingast, Barry 5, 130
'we-mode' 12, 115, 183
West, Stuart 6, 136
Westermarck, Edward 127
    effect 126
Wheeler, Peter 104
Whiten, Andrew 108, 145
Wiebe, Donald 35, 36, 37
Wilberforce, William 207
Wilkinson, Gerald 136
Wilson, David Sloan 38, 44, 47
Win Stay, Lose Shift 141
Wolf, Arthur 127
Wollstonecraft, Mary 168, 208
women's rights 167, 208
Woolcock, Michael 199
Wrangham, Richard 111, 120